Intraocular Lens Power Calculations

AVOIDING THE ERRORS

By

H. John Shammas, M.D.

Clinical Professor of Ophthalmology
USC School of Medicine, Los Angeles, California

Medical Director
M/S Surgery Eye Center, Lynwood, California

Intraocular Lens Power Calculations: Avoiding the Errors

H. JOHN SHAMMAS, M.D.
Shammas - Milkie Eye Medical Center
3510 M.L. King, Jr. Blvd.
Lynwood, CA 90262, USA

Library of Congress Catalog Card Number: 95-074935
ISBN: 0-915652-17-X
The News Circle Publishing House
Glendale, California

Printed in the United States of America
First Print 1996

Other books by H. John Shammas, M.D.
- *Atlas Of Ophthalmic Ultrasonography And Biometry* ISBN: 0-8016-4546-8
- *Color Atlas Of Ophthalmic Surgery: Cataracts* ISBN: 0-397-51071-3

Dedication

To my wife and best friend,
Najwa Mirhij-Shammas, M.D.
for her love, understanding and encouragement.

And to my three lovely daughters,
Rania, Maya, and Lina

TABLE OF CONTENTS

PREFACE

Intraocular lens (IOL) power calculations have become an integral part of the preoperative cataract evaluation. During the past decade, formulas used for such calculations have evolved to a high level of accuracy. The formulas are discussed in the first three chapters with their advantages and their limitations. Axial length measurement is one of the major factors affecting IOL power calculations. Biometry units are described in Chapter IV, stressing the clinical importance of the different elements within the biometer; Chapter V deals with the measurement of the axial length and Chapter VI with how to avoid errors in these measurements. Chapter VII discusses the corneal power and Chapter VIII the different constants required for the formulas.

A perfect cataract surgery has to go beyond the surgical technique; it has to meet the patient's visual needs and expectations. Chapter IX discusses IOL power selection in emmetropes, myopes and hyperopes. An error in IOL power calculations is usually suspected when the patient presents after the surgery with an unexpected induced myopia, hyperopia, or aniseikonia. In certain cases, the error is quite large necessitating a lens exchange as soon as possible. In Chapter X, I present a series of cases where errors in IOL power calculations have occurred, followed in each case by a discussion on the different alternatives for management.

IOL formulas will always be limited by the accuracy of the data used. This book is aimed at cataract surgeons and their technicians, to increase their understanding of biometry and IOL power calculations so that hopefully, major errors can be avoided.

In my professional life as an ophthalmologist, I have had two major areas of interest; the first one is "Ophthalmic Ultrasonography" and in 1984, I wrote a textbook on the subject, published by Mosby Co. My second area of interest is "Cataract Surgery" and how to improve its outcome. In 1991, I co-authored an "Atlas of Cataract Surgery" with Richard Kratz, M.D. The combination of these two areas of interest can only lead to the subject of biometry and IOL power calculations. This book represents the author's personal experience during the past 20 years.

H. John Shammas, M.D.
Clinical Professor of Ophthalmology
University of Southern California

Intraocular Lens Power Calculations

Avoiding the Errors

CHAPTER I

HISTORIC OVERVIEW

Before 1975 the power of an intraocular lens (IOL) to be inserted after a cataract extraction was calculated with the use of an equation based on the clinical history (1):

$$P = 18 + (1.25 \times Ref)$$

Where "P" is the power of an iris-supported intraocular lens for emmetropia, "Ref" is the preoperative refractive error in diopters before the development of the cataract. Errors exceeding 1 diopter occurred in over 50% of the cases with this clinical history method, and some errors were so large that they were referred to as the "9-diopter surprise." These large errors were caused by the difficulty of determining the patient's refractive error before the development of the cataract, and the large variations in the crystalline lens power.

A number of formulas for IOL power calculations have since been published; all these formulas are based on an accurate measurement of the corneal power and of the axial length. Original formulas were developed prior to 1980. They included theoretical formulas and regression formulas. All these formulas were then modified in the mid-1980's to correct some of the errors in the short and long eyes. Although these formulas are rarely used these days, they are at the basis of all modern formulas.

1. ORIGINAL THEORETICAL FORMULAS

All theoretical formulas for intraocular lens (IOL) power calculations are based on a two-lens system, the cornea and the pseudophakos lens, focusing images on the retina (1).

Required measurements

The required measurements for calculating the emmetropic lens power (P) and the expected postoperative refractive error (E) are:

(L): The axial length of the eye in millimeters, defined as the distance between the anterior surface of the cornea and the anterior surface of the retina

(C): The estimated postoperative anterior chamber depth in millimeters, defined as the distance between the anterior surface of the cornea and the anterior surface of the pseudophakos lens

(K): The corneal power in diopters or,

(R): The corneal radius of curvature in millimeters

Basic theoretical formulas for emmetropia

Thijssen's(2), Colenbrander's(3), Fyodorov's(4) and van der Heijde's(5) formulas yield approximately the same IOL power for emmetropia. Binkhorst's formula(6) differed by 0.50 diopter because it was based on a corneal index of refraction of 4/3 instead of 1.3375.

Thijssen's formula

$$P = \frac{1366}{L - C + \text{Const } 1} - \frac{1366}{\frac{1366}{K} - C + \text{Const } 2}$$

and where $\text{Const1} = \dfrac{\text{LENS THICKNESS X INDEX OF REFRACTION OF VITREOUS}}{\text{INDEX OF REFRACTION OF LENS}}$

Colenbrander's formula

$$P = \frac{1336}{L - C - 0.05} - \frac{1336}{\frac{1336}{K} - C - 0.05}$$

Fyodorov's formula

$$P = \frac{1336 - LK}{(L - C)\left(1 - \frac{CK}{1336}\right)}$$

van der Heijde's formula

$$P = \frac{1336}{L - C} - \frac{1}{\frac{1}{K} - \frac{C}{1336}}$$

Binkhorst's formula

$$P = \frac{1336\ (4R - L)}{(L - C)\ (4R - C)}$$

Resultant refractive error

The resultant postoperative refractive error(E) measured at the cornea after the insertion of an IOL with a certain power(I) can be predicted before surgery by Binkhorst's formula(6) or Hoffer's modification of Colenbrander's formula(7):

Hoffer's formula
$$E = \frac{1.336}{\frac{1.336}{\frac{1336}{L - C - 0.05} - I} + \frac{C + 0.05}{1000}} - K$$

Binkhorst's formula
$$E = \frac{1336\ (4R - L)\ -\ I\ (L - C)\ (4R - C)}{1336\ (0.003\ LR) - I\ (L - C)\ (0.003\ CR)}$$

2. ORIGINAL REGRESSION FORMULAS

Regression formulas are derived empirically from retrospective computer analysis of data on a great many patients who have undergone surgery.

Required measurements

The required measurements for calculating the emmetropic lens power(P) and the expected postoperative refractive error(E) are:

(L): The axial length of the eye in millimeters, defined as the distance between the anterior surface of the cornea and the anterior surface of the retina

(K): The corneal power in diopters

Regression formula for emmetropia

A regression formula for emmetropia is based on the following equation:

$$P = A - BL - CK$$

where P is the IOL power for emmetropia, L is the axial length in millimeters and K is the corneal power in diopters. A, B, C are constants.

The SRK formula(8,9) is the most popular original regression formula where B is 2.5 and C is 0.9. The formula becomes:

$$P = A - 2.5L - 0.9K$$

The constant A varies with the IOL style and manufacture. Different values for the constant A are discussed in Chapter VIII.

Resultant refractive error

The formula(8) is: $E = 0.67\ (P - I)$

where E is the resultant refractive error, P is the power of the IOL for emmetropia, and I the power of the inserted IOL. The formula is based on data analysis of over 2500 lenses. A change of 1.50 diopter implant power produces a change of approximately 1.00 D in final postoperative refraction.

Comparing the SRK regression formula to the original theoretical formulas

Figure 1 shows the difference in the calculated power for emmetropia between the original formulas and the SRK regression formula. The values are almost identical when the axial length ranges between 23 and 24 mm. The SRK formula measures a stronger IOL power for emmetropia in the long eyes and a much weaker IOL power in the short eyes.

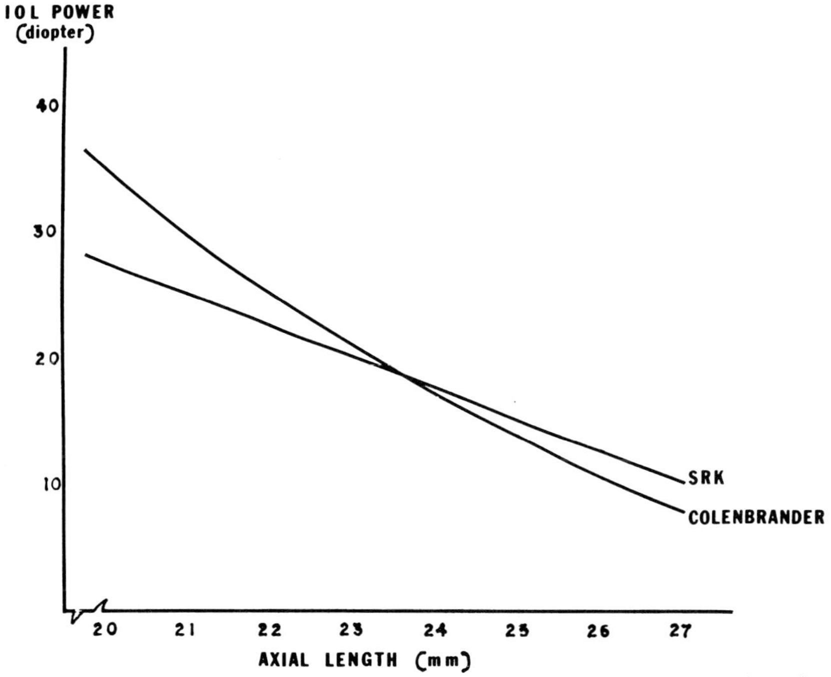

Figure 1: Calculated IOL power by the SRK and Colenbrander formulas for axial lengths ranging between 20 and 27mm.

3. MODIFIED FORMULAS FOR EMMETROPIA

These formulas are modifications of the original theoretical and regression formulas to correct for errors occurring in the long and short eyes.

Hoffer's formula

Hoffer's formula(10) is based on a modification of Colenbrander's formula. The expected post-operative error at the cornea(R) is added to the corneal power(K). When calculations are made for emmetropia, R becomes nil.

Also, the anterior chamber depth constant has to be changed in accordance with the axial length:

$$\text{corrected } C = (0.292\,L - 2.93) + (C - 3.94)$$

The Hoffer formula is written:

$$P = \frac{1336}{L - C - 0.05} - \frac{1.336}{\dfrac{1.336}{K + R} - \dfrac{C + 0.05}{1000}}$$

Where:

L = axial length (mm)

C = anterior chamber depth (mm)

K = corneal power (D)

P = IOL power (D)

R = refractive error (D) at the cornea

Shammas' formula

Shammas' formula(11) is a modified Colenbrander's formula. The first modification involves the corneal index of refraction; it has been changed from 1.3375 to 4/3, as suggested by Binkhorst. The corneal power, such as read on the Bausch and Lomb Keratometer, is decreased by a factor of 1.0125. The second modification is the incorporation of a fudge factor linked to the axial length. This fudge factor has been established by reviewing 500 cases of axial length measurements and IOL power calculations in which this factor was needed for higher accuracy. These cases were divided into six categories according to the axial length; there was no difference between the calculations and final refractions when the axial length was 23 mm. The average difference was +0.25D for each mm shorter than 23 mm and -0.25D for each mm longer than 23 mm. Increasing the axial length by 0.1 mm for each 1 mm shorter than 23 mm, and decreasing it by 0.1 mm for each 1 mm longer than 23 mm, closes the gap between the calculations and the final refractions in short and long eyes. This fudge factor can be introduced easily into any theoretical formula by changing the axial length(L) to L - 0.1 (L - 23).

The Shammas formula becomes:

$$P = \frac{1336}{L - 0.1\,(L - 23) - C - 0.05} - \frac{1}{\frac{1.0125}{K} - \frac{C + 0.05}{1336}}$$

Where:

P = power of the IOL for emmetropia in diopters

L = axial length in millimeters

K = average keratometer readings in diopters

C = distance between the anterior vertex of the cornea and the IOL, in millimeters (estimated anterior chamber depth)

The Shammas formula is based on axial length measurements taken through an immersion technique and does not add any retinal correction. However, in cases where the axial length is measured by a contact technique, 0.24 mm should be added to the measurement.

Binkhorst's adjusted formula

Binkhorst (12) used his own original formula for emmetropia and varied the post-operative anterior chamber depth according to the axial length.

$$P = \frac{1336\,(4R - L)}{(L - C)\,(4R - C)}$$

Where:

P = power of the IOL for emmetropia in diopters
L = axial length in millimeters
R = radius of curvature of the anterior surface of the cornea in millimeters
C = distance between the anterior vertex of the cornea and the IOL in millimeters (estimated anterior chamber depth).

The postoperative anterior chamber depth estimated values (C) are adjusted for iris-fixated, iridocapsular, and posterior chamber lenses. This is based on a positive correlation noted in these cases by Binkhorst between the postoperative anterior chamber depth (C) and the axial length (L).

In the presence of a posterior chamber lens, C is decreased by 0.17mm for each millimeter the axial length is shorter than 23.45mm; C is increased by 0.17mm for each millimeter the axial length is longer than 23.45mm.

The SRK II formula

The SRK II formula(13) is a modification of the original SRK formula:

P = A - 2.5L - 0.9K + correction

Where:

P = implant power in diopters to produce emmetropia
L = axial length in mm
K = average keratometric readings in diopters
A = the specific constant for each lens type and/or manufac-turer.

In eyes with an axial length ranging between 22.0mm and 24.5mm, there is no change in the SRK formula, and the correction factor is not used.

In the short eyes, measuring less than 22.0mm, the SRK formula predicts a lower emmetropic IOL power than is actually true. It is corrected by:

- adding 1 diopter to the calculated P value, if the axial length ranges between 21.0mm and 21.9mm

- adding 2 diopters to the calculated P value, if the axial length ranges between 20 and 20.9mm

- adding 3 diopters to the calculated P value, if the axial length is under 20 mm.

In the long eyes, measuring over 24.5mm, the SRK formula predicts a slightly higher emmetropic IOL power than is actually true. It is corrected by subtracting 0.5 diopters from the calculated P value.

The SRK formula for ametropic calculations has been rewritten to read:

$$I = P - (Rt \times Rf)$$

Where:

I = the implant power in diopters for a target refraction
P = the implant power in diopters for emmetropia
Rt = the desired postoperative refraction and
Rf = the refractive factor

In the SRK formula, Rf is 1.25 if the power for emmetropia exceeds 14 D and 1.0 if the power for emmetropia is 14 D or less.

Other modifications of the SRK formula have been developed during the past decade to improve its accuracy. Some of the best known are:

- The Thompson, Maumenee and Baker formulas(14), and the,
- Donzis, Kastl and Gordon formula(15).

REFERENCES - CHAPTER I

1. Shammas, HJ: Atlas of Ophthalmic Ultrasonography and Biometry. pp. 273-308. The C.V. Mosby Co., St. Louis, 1984.

2. Thijssen, JM: The emmetropic and iseikonic implant lens: computer calculation of the refractive power and its accuracy. Ophthalmologica. 171:467-486, 1975.

3. Colenbrander, MC: Calculations of the power of an iris clip lens for distance vision, BR.J. Ophthalmol. 57:735-740, 1973.

4. Fyodorov, SN, Galin, MA, and Linksz, A: calculation of the optical power of intraocular lens, Invest. Ophthalmol. 14:625-628,1975.

5. van der Heijde, GL: The optical correction of unilateral aphakia, Trans. Am. Acad. Ophthalmol. Otolaryngol. 81:80-88, 1976

6. Binkhorst, RD: The optical design of intraocular lens implants, Ophthalmic Surg. 6:17-31, 1975.

7. Hoffer, KJ: Accuracy of ultrasound intraocular lens calculation, Arch. Ophthalmol. 99:1819-1823, 1981.

8. Sanders, DR, Kraff, MC: Improvement of intraocular lens power calculation using empirical data. Amer. Intraoc. Implant Soc.J. 6:263-267, 1980.

9. Sanders, DR, Retzlaff, J, Kraff, MC: Comparison of empirically derived and theoretical aphakic refraction formulas. Arch. Ophthalmol. 101:965-967, 1983.

10. Hoffer, KJ: Intraocular lens calculations: the problem of the short eye. Ophthalmic Surg. 12:269-272, 1981.

11. Shammas, HJF: The fudged formula for intraocular lens power calculations. Am. Intraoc. Implant Soc. J. 8:350-352, 1982.

12. Binkhorst, RD: Intraocular Lens Power Calculation Manual: A Guide to The Author's TICC-40 Programs, ED 3, 1984, New York, R.D. Binkhorst.

13. Sanders, DR, Retzlaff, J, Kraff, MC: Comparison of the SRK II formula and the other second generation formulas. J. Cataract Refract. Surg. 14:136-141, 1988.

14. Thompson, JT, Maumenee, AE, Baker, CC: A new posterior chamber intraocular lens formula for axial myopes. Ophthal. 91:484-488, 1984.

15. Donzis, PB, Kastl, PR, Gordon, RA: An intraocular lens formula for short, normal and long eyes. Cont. Lens Ass. Ophthalmol. J., 11:95-98,1985.

CHAPTER II

CALCULATING FOR EMMETROPIA

Modern theoretical formulas are more complex than the original and modified formulas. The most striking difference is the manner with which the estimated anterior chamber depth is calculated:

- In the original formulas, the anterior chamber depth is a constant value.
- In the modified formulas, the anterior chamber depth varies with the axial length. It decreases in the shorter eye and increases in the longer eye.
- In the modern formulas, the anterior chamber depth varies, not only with the axial length, but with the corneal curvature as well. The anterior chamber is deeper in the presence of a steep cornea and shallower in the presence of a flat cornea.

The best known modern formulas are the Holladay formula(1), the SRK-T formula(2) and the Hoffer-Q formula(3). Two additional formulas based on Gaussian optics were written by Olsen(4) and by Colliac(5). These two formulas are better known in Europe and are rarely used in the USA.

We will confine our discussion to the first three formulas. Each formula comprises multiple equations and, instead of listing these equations, we will relate the rationale of these formulas. These formulas are readily available on calculators, computer diskettes and ultrasound units.

1. HOLLADAY'S FORMULA

Holladay's formula (1) is a modification of the theoretical formula based on a three-part system that includes data screening criteria to identify improbable axial length and keratometry measurements, a more accurate postoperative anterior chamber depth estimate that increases the accuracy in short,medium and long eyes, and a personalized "surgeon factor" that adjusts for any consistent bias in the surgeon's formula.

The data screening criteria

The data screening identifies measurements that are unusual and might require remeasurement such as:

- axial length less than 22.0 mm or over 25.0 mm
- average corneal power less than 40 diopters or over 47 diopters.
- calculated emmetropic IOL power of more than 3 diopters from specific lens style.
- a difference between the two eyes of over 1 diopter in the average corneal power, of over 0.3 mm in the axial length and of over 1 diopter in the emmetropic IOL power.

Measurement of the postoperative anterior chamber depth estimate.

The effective lens position of any IOL is the sum of the anatomic ACD and the distance from the anterior plane of the iris to the optical plane of the IOL, called the surgeon factor(S).

A personalized surgeon factor

The surgeon factor is calculated by solving the formula in reverse, using as input variables the postoperative corneal and axial length measurements, the IOL power implanted and the stabilized postoperative refraction.

2. THE SRK-T FORMULA

The SRK-T formula (2), contrary to the SRK I and the SRK II formulas, is a theoretical formula based on Fyodorov's formula and uses empirical regression methodology for optimization of the post-operative ACD prediction, of the retinal thickness correction factor and of the corneal refractive index.

Optimization of the postoperative ACD prediction

The estimated postoperative ACD is calculated:

ACD (estimate) = corneal height (H) + offset
offset = ACD (constant) - 3.336

and is based on the hypothesis that the IOL lies at a constant distance from the calculated iris plane.

The height of the corneal dome (H) is the distance between the cornea and the iris-plane and is calculated mathematically within the formula.

Axial length correction factor for retinal thickness

The optical axial length (LOPT) is measured by adding the retinal thickness (RETHICK) to the axial length (L)

LOPT = L + RETHICK

The corneal refractive index

The corneal radius of curvature in mm is measured: r = 337.5/K where K is the average keratometric reading in diopters.

3. HOFFER-Q FORMULA

The Hoffer-Q formula(3) uses the basic Hoffer's modification of Colenbrander's formula with a new anterior chamber depth (ACD) prediction formula. Hoffer studied the relationship between the ACD and the axial length and found it to be a tangent curve instead of a straight line. He then tried, by trial and error, many mathematical formula variations until the desired curve was reached.

The anterior chamber prediction formula consists of:

- A personalized ACD developed from any series of one IOL style.
- A factor that increased the ACD with increasing axial length.
- A factor that increased the ACD with increasing corneal curvature.
- A factor that moderated the change in ACD for extremely long (over 26 mm) and short (less than 22 mm) eyes, and
- A constant added to the ACD.

4. COMPARING THE FORMULAS

Figure 2 shows that the results of the modified modern formulas fall between the results of the original theoretical (Colenbrander) and regression (SRK) formulas.

The following table shows that all modern modified formulas yield approximately the same results in an average eye (A.L. of 23.5 mm and K readings of 43.5 diopters), in a short eye with a steep cornea (A.L. of 21.0 mm and K readings of 47.0 diopters) and in a long eye with a flat cornea (A.L. of 26.0 mm and K readings of 41.0 diopters).

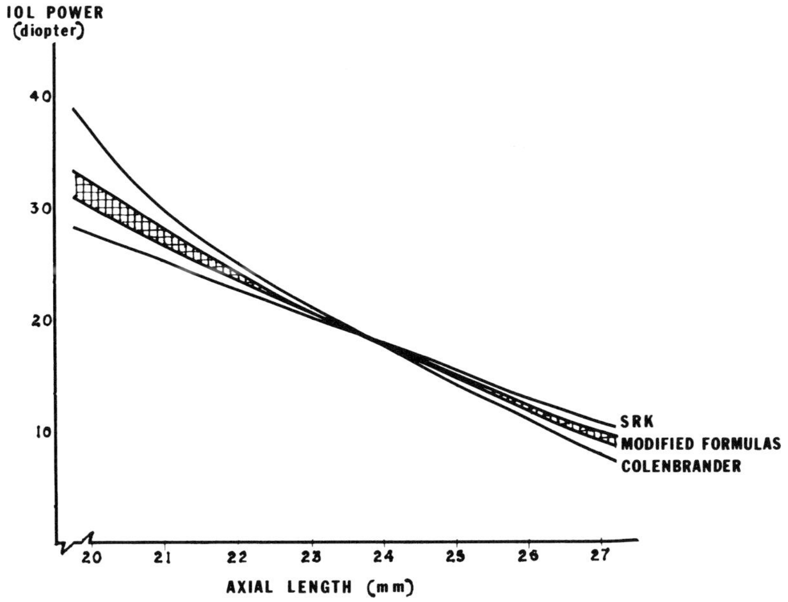

Figure 2: Calculated IOL power by the modified modern formulas (shaded' areas) compared to the calculated powers by the SRK and Colenbrander's formulas.

IOL POWER FOR EMMETROPIA

Axial length (mm)	21.0	23.5	26.0
K readings (D.)	47.0	43.5	41.0
SHAMMAS	23.7	19.3	15.4
SRK - T	23.8	19.3	15.4
HOLLADAY	23.7	19.3	15.1
HOFFER-Q	23.9	19.3	15.3
BINKHORST	23.7	19.3	15.3

Variations between the modern formulas still exist, especially in the short and long eyes. These variations seem to be related to changes in the corneal curvature. Figure 3 shows that in a 21 mm eye, the formulas differ by almost two diopters in the unusual short eyes with flat corneas (41 diopters) and only by 0.2 D. in the most commonly encountered short eyes with steeper corneas (47 diopters).

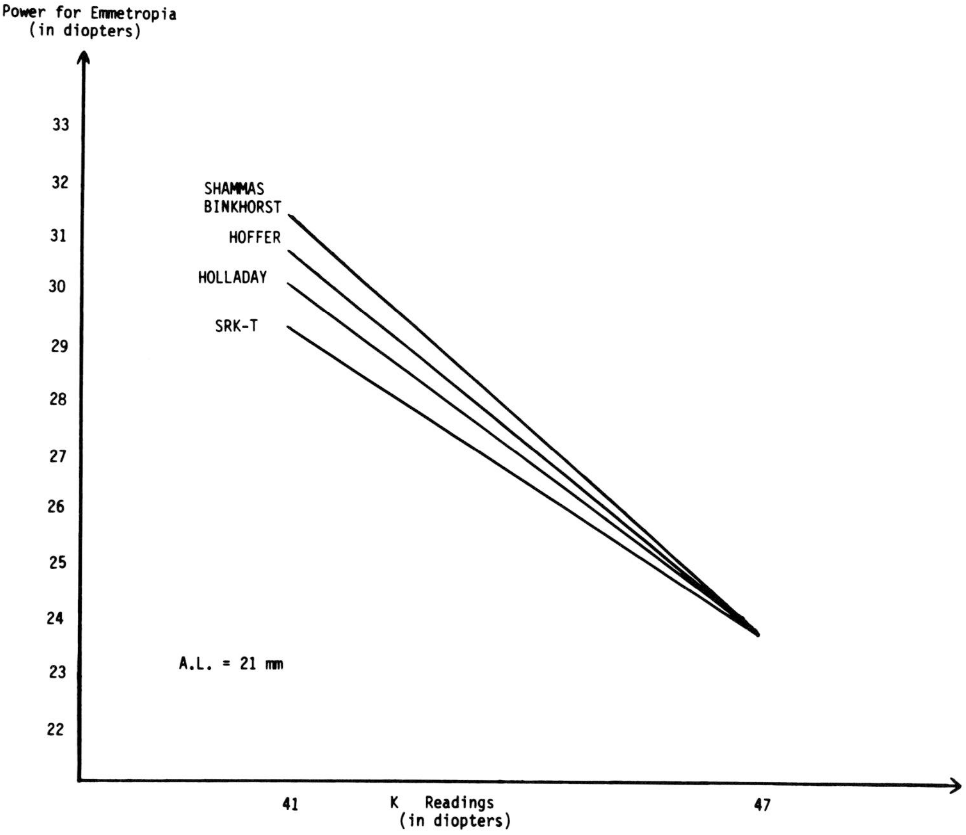

Figure 3: Variations in the calculated IOL power for emmetropia with the corneal curvature in the short 21mm eye.

Similarly in the 26 mm long eye (Figure 4), the formulas differ by almost 1.3 diopters in the unusual long eyes with steep corneas (47 diopters) and differ by less than 0.5 diopters in the most common long eyes with a flat cornea (41 diopters).

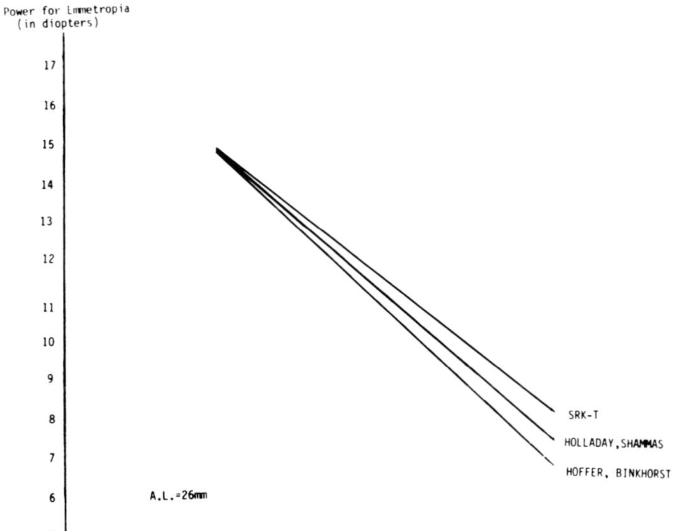

Figure 4: Variations in the calculated IOL power for emmetropia with the corneal curvature in the long 26mm eye.

5. CLINICAL RESULTS

It was once said that each author usually finds his formula to be the most accurate ,well, at least in his series, and that's why each formula is personalized to account for the variations in the biometry unit, the axial length measurement, the specific keratometer, the implant being used and even the surgical technique.

The postoperative results in 230 consecutive cases of extracapsular cataract extraction and intraocular lens implant surgery were reviewed. All lenses used were plano-convex posterior chamber lenses, model SK62CP from Cilco, Inc. with 10 degree angle loops. The axial length was measured with the Kretz 7200MA standardized A-Scan, using an immersion technique. The K readings were taken with a Bausch and Lomb keratometer. All postoperative K readings were within +/- 0.50 diopter of the preoperative readings(6).

Six formulas were used to calculate the IOL power for emmetropia. They included the Shammas, SRK II, SRK-T, Holladay, Hoffer-Q and Binkhorst II formulas. A specific constant was calculated for each formula. These values were established by reviewing the results of 20 surgeries on eyes with an axial length ranging between 23.0 and 24.0mm, and K readings ranging between 43.0 and 44.0 diopters. The constant was adjusted for each formula until the final refraction equaled the predicted refraction in each eye. The values were then averaged to obtain the personalized constant for each formula. The following constants were obtained: anterior chamber depth constants (ACD) of 3.90 with the Shammas formula; 4.46 with the Binkhorst II formula; 4.50 with the Hoffer-Q formula; an A constant of 117.3 for the SRK-T and SRK II formulas; and an S factor of 0.74 for Holladay formula. The difference between the final and predicted refractions was tabulated for each eye and for each of the six formulas. The mean absolute error, the range, and the accuracy obtained with these formulas were as follows:

MODERN FORMULAS

Formulas	Mean Absolute Error In Diopters	Range Of Errors In Diopters
Shammas formula	0.52 +/- 0.37	+1.3 to -1.6
SRK-T formula	0.54 +/- 0.39	+1.4 to -1.7
Holladay formula	0.52 +/- 0.39	+1.4 to -1.6
Hoffer-Q formula	0.54 +/- 0.39	+1.4 to -1.6
Binkhorst II formula	0.52 +/- 0.38	+1.3 to -1.6
SRK II formula	0.65 +/- 0.42	+2.4 to -2.3

ACCURACY WITHIN +/- 1 DIOPTER

Formula	All Cases (n=230)	Axial Length <23mm (n=70)	Axial Length 23 - 24mm (n=97)	Axial Length >24mm (n=63)
Shammas	220(95.7%)	64(91.4%)	94(96.9%)	62(98.4%)
SRK-T	218(94.8%)	62(88.6%)	95(97.9%)	61(96.8%)
Holladay	223(97.0%)	65(92.9%)	96(99.0%)	62(98.4%)
Hoffer-Q	219(95.2%)	64(91.4%)	94(96.9%)	61(96.8%)
Binkhorst II	220(95.7%)	64(91.4%)	94(96.9%)	62(98.4%)
SRK II	210(91.3%)	56(80.0%)	95(97.9%)	59(93.7%)

Conclusions:

. All modern theoretical formulas are quite accurate for predicting the IOL power for emmetropia

. The SRK II formula performed well in the average eyes but did not perform as well in the short eyes. It should be used with caution. By the same token, the SRK-I has been proven to be even less accurate in the short eyes.

REFERENCES - CHAPTER II

1. Holladay, JT, Praeger, TC, Chandler, TY,Musgrove, KH: A three-part system for refining intraocular lens power calculations. J. Cataract Refract Surg. 14:17-24, 1988.

2 . Retzlaff, J, Sanders, DR, Kraff, MC: Development of the SRK/T intraocular lens implant power calculation formula. J. Cataract Refractive Surg 16:333-340, 1990.

3. Hoffer, KJ: The Hoffer-Q formula: A comparison of theoretic and regression formulas. J. Cat. Refract. Surg. 19:700-712, 1993.

4. Olsen, T, Thim, K, Corydon, L: Theoretical versus SRK I and SRK II calculation of intraocular lens power. J. Cataract Refractive Surg. 16:217-225, 1990.

5. Colliac, JP: Matrix formula for intraocular lens power calculation. Investigative Ophthalmol. & Visual Sc. 31:374-381, 1990.

6. Shammas, HJ: Accuracy of the modified IOL power formulas for emmetropia. Middle East J. Ophthalmol. 1:33-37, 1993.

CHAPTER III

CALCULATING FOR ISEIKONIA

Calculating for iseikonia becomes important if the eyes are highly myopic or highly hyperopic, with a cataract in one eye necessitating surgery while the fellow eye maintains good vision and does not require any surgery.

If iseikonia with the fellow eye is not restored, the two retinal images of the same object become of unequal size. This is called Aniseikonia. It is manifested mainly by asthenopia or rapid fatigue of the eyes accompanied with headaches. Other symptoms include nervousness, vertigo, dizziness and general fatigue.

Mathematically, aniseikonia is calculated by the formula:

$$\text{Aniseikonia (\%)} = \frac{f1 - f2}{f2} \times 100$$

Where f1 is the posterior focal length of the phakic fellow eye and f2 the posterior focal length of the pseudophakic eye. Clinically, the eyes can tolerate up to 5% in aniseikonia. This is why we introduced the concept of the "5% iseikonic range". Practically, this range represents a variation of +/- 2.50 diopters from the expected refraction when the iseikonic lens is used.

1. IOL FORMULAS FOR ISEIKONIA

Multiple formulas have been written to calculate the iseikonic lens implant. Some of the most popular formulas are herein described.

R.D. Binkhorst's formula

The iseikonic IOL power (D) is calculated with the following formula(1,2):

$$D = \frac{1336 \quad (3\,r/F1) - 1}{4r - d}$$

where r is the radius of curvature of the corneal anterior surface and d is the anterior chamber depth of the eye.

F1 is the anterior focal length of the fellow eye in mm and is calculated mathematically within the formula.

van der Heijde's formulas

These formulas (3,4) are based on the principle that for distant objects, the size of the optical image is proportional to the posterior focal length of the total system eye plus correction. Nomograms also are available for quick reference. These formulas are rewritten by Huber and Binkhorst(5):

$$FI = \frac{n}{L - d_2 - 6.2\,(10^{-5})} - \frac{1}{\frac{1}{FC'} - \frac{d2}{n}}$$

$$Fc' = Fc + \frac{1}{\frac{1}{F_s} - d_1}$$

Where:

$d1$ = Spectacle-to-corneal vertex distance (12mm)
$d2$ = Anterior segment depth
Fs = Spectacle power
Fc = Corneal power
Fc' = Corrected corneal power
$F1$ = Power of intraocular lens (plano-convex iridocapsular lens, 0.6 mm thick)
L = Axial length of the eye
n = Refractive index of aqueous and vitreous = 1.336

Hoffer's iseikonic formula

Hoffer iseikonic formula (6) is written:

$$I = \frac{1336}{L - C - 0.05} - \frac{1.336}{\frac{1.336}{K+S} - \frac{C+0.05}{1000}}$$

Where:

I = iseikonic IOL power in diopters
L = axial length of opposite eye in mm (minus 0.657 mm if phakic only)
C = anterior chamber depths in mm
K = corneal power in diopters
S = spherical equivalent refractive error in diopters of opposite eye

In most cases, the operated eye needs to be made two diopters more myopic to obtain iseikonia. This is true when an iris-supported or an anterior chamber lens is used, but the formula cannot be used for posterior chamber lenses.

2. SHAMMAS ISEIKONIC EQUATIONS

The IOL power for iseikonia(Pi) is calculated using the same formula for emmetropia with modifications in the values of the axial length and of the K readings(7).

Calculating the iseikonic axial length

In this formula the axial length L is substituted by Li where:

$$Li = Lopp - \frac{1}{3}(5.85 - C)$$

Where Lopp is the axial length of the opposite phakic eye and C is the estimated post-operative anterior chamber depth.

These changes in the axial length are based on calculations done with van der Heijde's nomograms. The concept is to keep the posterior focal length equal in both eyes in order to have the same image size. Theoretically in an average eye the lens can be assimilated to a thin lens at 5.85 mm behind the corneal vertex. When an implant is inserted it is located between 2.9 mm (anterior chamber lens) and 5.0 mm (biconvex posterior chamber lens) behind the corneal vertex. This creates a certain elongation of the posterior focal length of the implanted eye if an emmetropic IOL is used. In order to keep the same posterior focal length as the fellow phakic eye the iseikonic implant has to be stronger and focus the image in front of the retina, thus making the implanted eye more myopic. In normal cases, the required induced myopia in the pseudophakic eye in order to keep the same image size as in the fellow phakic eye varies with the type of implant and where it sits in the eye:

Implant	AC depth	Induced myopia
Anterior chamber lens	2.9 mm	- 2.50 D
Iris supported lens	3.5 mm	- 2.00 D
Plano-convex PC lens	4.3 mm	- 1.25 D
Biconvex PC lens	5.0 mm	- 0.75 D
Biconvex PC lens	5.3 mm	- 0.25 D

The axial length measurement is not changed if the fellow eye is already pseudophakic since both implants will be almost at the same distance from the corneal apex.

Calculating the iseikonic corneal power

The second modification to the formula concerns the corneal power (K). It is changed to Ki where:

$$Ki = K + R\ opp$$

R opp is the refraction of the opposite eye at the cornea. For refractions of - 3.0 to + 3.0, this number is almost the same as the spectacle refraction. In the presence of a higher refraction, the refraction at the cornea (R) can be calculated from the spectacle correction (R spec)

$$R = R\ spec \times \frac{1}{1 - 0.012\ R\ spec}$$

Calculating the iseikonic IOL power

The two equations involving the changes in the axial length and in the K readings can be incorporated into any IOL power formula to derive the iseikonic IOL power and the final refraction achieved with such an IOL.

Case Presentation

A 71-year-old patient presents with a cataract and decrease vision in her left eye. The right eye only shows incipient cataract changes. The preoperative measurement are as follows:

	OD	OS
Vision	20/40	20/200
Refraction	-6.50 Sph	-6.50 -0.75X84
Axial length	26.85 mm	26.90 mm
K readings	45.25/45.25 D	45.50/45.50 D
IOL for emmetropia	8.04 D	7.58 D

The decision is to operate on the left cataractous eye and, since the right eye is still phakic and does not require surgery in the near future, it becomes imperative to calculate:

- -The iseikonic lens power
- - The final refraction that will be achieved with such an IOL
- - And the 5% tolerance range

First we calculate Li:

$$Li = L\ opp - (1/3)\ (5.85 - C)$$

$$Li = 26.85 - (1/3)\ (5.85 - 4.50) = 26.40\text{mm}$$

Then we calculate Ki:

$$Ki = K + R\ opp$$

$$Ki = 45.50 - \frac{6.50}{1 - (0.012)(-6.50)} = 39.50 \text{ diopters}$$

The IOL power for iseikonia is calculated with an axial length (Li) of 26.4 mm, and K readings (Ki) of 39.50 diopters.

The IOL power for iseikonia is calculated to be 17.0 diopters.

The final refraction achieved with such an IOL is calculated to be -7.25 Sph.

The 5% iseikonic range (approximately +/- 2.50 diopters) will then be from -4.50 Sph to -10.00 Sph. It is always advisable to leave the operated eye with the least myopia possible, without inducing any aniseikonia. In this specific case, the final refraction should then be around -4.50 Sph to avoid any aniseikonia. At the same time, you also would have avoided any significant anisometropia between the two eyes.

The IOL calculations are then repeated for the left eye using the measured axial length for that eye (26.90 mm) and the measured K readings (45.50 diopters). The IOL needed to produce a final refraction of -4.50 Sph is calculated to be 13.2 diopters.

REFERENCES - CHAPTER III

1. Binkhorst, RD: The optical design of intraocular lens implants, Ophthalmic Surg. 6:17-31, 1975

2. Binkhorst, RD: Intraocular Lens Power Calculation Manual: A Guide to The Author's TICC-40 Programs, ED 3, 1984, New York, R.D. Binkhorst

3. van der Heijde, GL: A nomogram for calculating the power of the pre-pupillary lens in the aphakic eye, Bibl. Ophthalmol. 83:273-275, 1975.

4. van der Heijde, GL: The optical correction of unilateral aphakia, Trans. Am. Acad. Ophthalmol. Otolaryngol. 81:80-88, 1976.

5. Huber, C and Binkhorst, C: Iseikonic lens implantation in anisometropia. Amer. Intra-Ocular Implant Soc. J. 5:194-198, 1979.

6. Hoffer, KJ: Intraocular lens calculations: the problem of the short eye. Ophthalmic Surgery. 12:269-272, 1981.

7. Shammas, HJ: Integrated iseikonic equation for intraocular lens power calculations. J.Cat. Refract. Surg. 20:74-77, 1994.

8. Troutman, RC: Artiphakia and aniseikonia. Trans. Am. Ophthalmol. Soc. 60:590-658, 1962.

CHAPTER IV

AXIAL LENGTH MEASUREMENT: THE BIOMETER

A-scan biometers are instruments used to measure the axial length. Most newer units are compact, efficient, computerized and complete with IOL power calculation capabilities, and a print-out of the measurements and of the results. Biometry also is provided on larger ultrasound units used for diagnostic B-scan examination, standardized A-scan echography and even pachymetry.

Measurement of the axial length is influenced by many elements within the biometer. This requires a better understanding of ultrasound physics and of the clinical importance of the different components of a biometer.

1. BASICS OF ULTRASOUND

Ultrasound consists of high frequency sound waves over 20,000 cycles per second, which is the highest frequency audible to the human ear.(1)

Ultrasound propagates within a medium in a longitudinal manner as alternate condensations and rarefactions. These pressure variations are known as oscillations. While spreading through a medium, oscillations are characterized by a velocity, a frequency and a wavelength. The relationship between these three factors is:

$$\text{velocity} = \text{wavelength} \times \text{frequency}$$

Frequency

The frequency(2) is the number of hertz (Hz) or cycles per second. Higher frequencies provide a higher resolution while lower frequencies provide better penetration, but a reduction in the resolution.

Clinical application:
To obtain the high resolution needed for axial length measurement, biometry units use ultrasound frequencies ranging between 8 and 15 MHz. (1 MHz = 1 megahertz = 1 million cycles per second).

Wavelength

The wavelength(3) is the distance between two particles in the same phase of oscillation. Within the ocular tissues, the wavelength is approximately 0.19 mm if an 8 MHz probe is used and 0.15 mm if a 10 MHz probe is used.

Velocity

The velocity(4) is the speed of sound propagation and is expressed in meters per second (m/s). The velocity varies according to the medium through which sound propagates; within the eye, the ultrasound velocity is 1532 m/s in aqueous and vitreous, 1641 m/s in a clear lens and 1550 m/s in solid tissues.

Clinical application:
During an accurate measurement of the different eye components, the proper sound speed has to be used for each of these entities.

Ultrasound beam

An ultrasound beam is emitted from the transducer and is directed toward the tissues under examination(5,6). During biometry, the emitted sound will meet multiple interfaces. At each interface, part of the sound beam is reflected back toward the probe and the remainder of the sound beam keeps propagating deeper into the tissues. This process will generate echospikes from the different interfaces that have been intersected, i.e., anterior surface of the cornea, posterior surface of the cornea, anterior surface of the lens, posterior surface of the lens, anterior surface of the retina and anterior surface of the sclera. When the ultrasound beam reaches the orbital tissues, it is attenuated until it loses all its energy. The sound beam returns to the transducer that also acts as a receiver. The pulses are then processed within the biometer to display "echo-signals" on the screen.

Clinical application:
During axial length measurement, it is important to display "maximal echo-spikes" by keeping the incident beam perpendicular to the different interfaces. Only then will the returning beam carry the maximal energy reflected by the tissues. This is especially true about the retinal spike:

-When the probe is aligned with the optical axis of the eye, the sound beam is perpendicular to the retinal interface. The portion of the returning beam that reaches the probe will carry maximal energy from the surface under examination (i.e. the retina) and the retinal spike is displayed as a straight, steeply rising echosignal (Fig. 5).

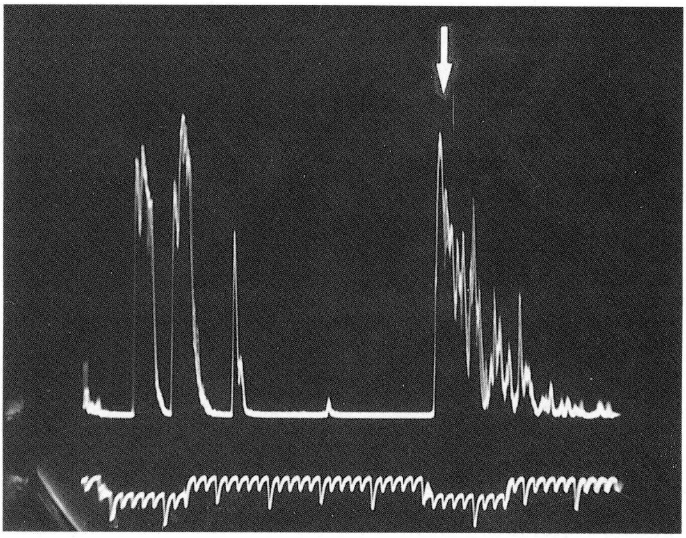

Figure 5: The retinal spike (arrow) is displayed as a straight, steeply-rising echospike.

-When the probe is not properly aligned with the optical axis of the eye, the sound beam becomes tangential to the retinal surface, instead of being perpendicular to it. The portion of the returning beam that reaches the probe will carry minimal energy from the surface under examination (i.e the retina) and the retinal spike is displayed as a jagged, slow-rising echospike (Fig. 6).

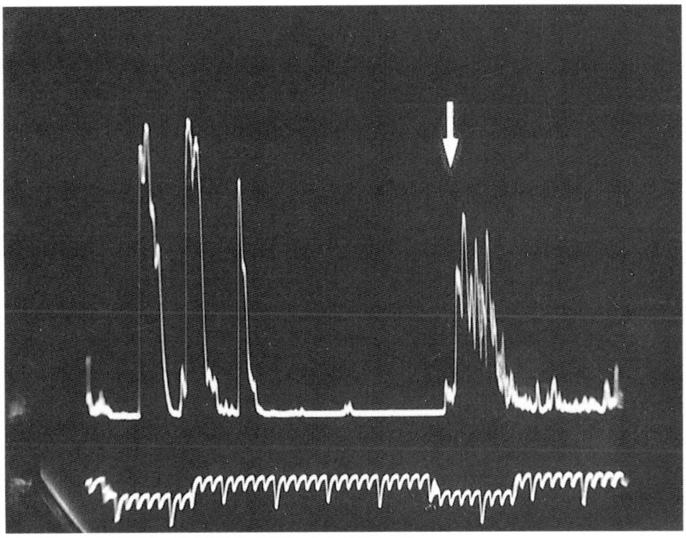

Figure 6: The retinal spike (arrow) is displayed as a jagged slow-rising echospike.

2. BIOMETRY UNITS

Biometry units vary from simple to highly sophisticated. Five different components are herein discussed because they have a direct bearing on the measurement. These are the probe configuration, the transducer, the oscilloscope, the sensitivity and velocity settings. The clinical importance of each component is also discussed.

The Probe

The probe is connected to the main chassis of the biometry unit by an electronically shielded cable and contains a transducer at its tip.

- The solid probe (Fig.7) such as the one available with the Kretz 7200 MA ultrasound unit is basically used to measure the axial length through an immersion technique. This probe is particularly popular with ultrasonographers who use it for standardized A-scan echography.
- The newer and thinner solid probes (Fig. 8 & 9) are becoming more popular for axial length measurement. They can be used for immersion technique in the same manner as the Kretz solid probes. Furthermore, when used for a contact technique, they produce minimal corneal indentation.

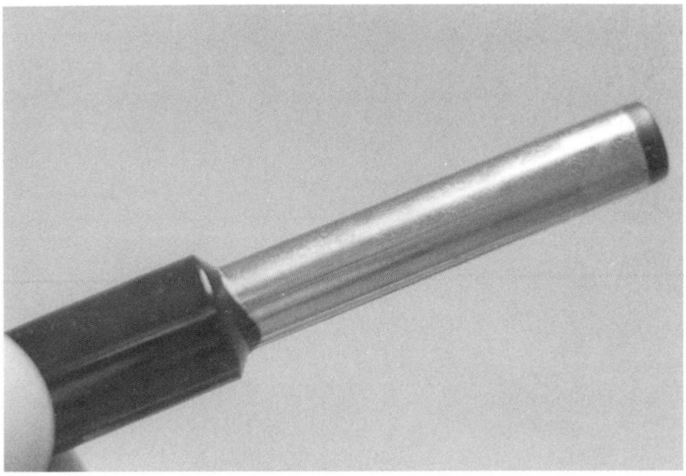

Figure 7: Solid probe available on the Kretz 7200 MA ultrasound unit.

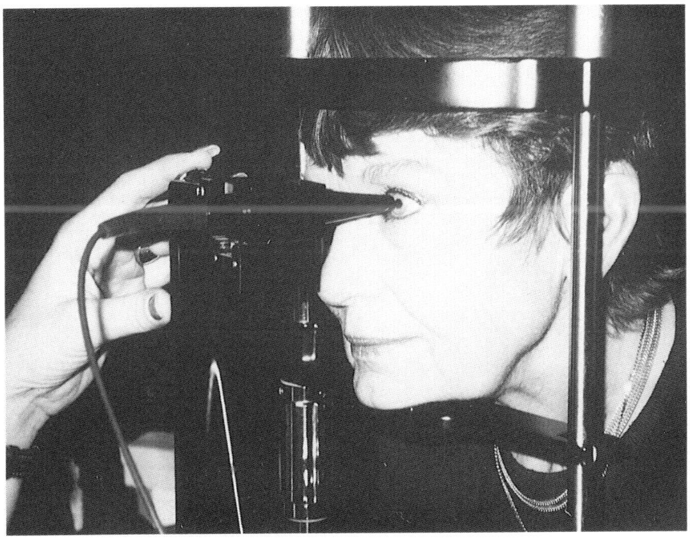

Figure 8: Thinner solid probe used with most newer biometry units.

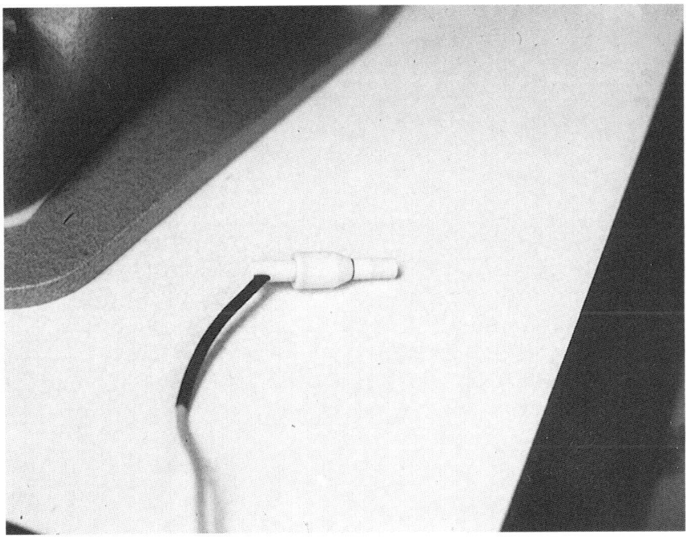

Figure 9: Small solid probe that attaches to the tonometer holder.

- The semi-soft probe (Fig.10) has a membrane at the tapered end of a cylinder fastened with an O-ring. The cylinder fits into a spring-loaded transducer carriage connected to a joystick manipulator. It allows the axial length to be measured through a contact technique. These semi-soft probes were once very popular, but they require a certain amount of maintenance such as a regular exchange of the membrane, the O-ring, and of the water within the cylinder. Many ultrasound companies have phased out these semi-soft probes and replaced them with the smaller, less cumbersome compact solid probes.

- The soft probe (Fig. 11) has a loose membrane at its tip. It is used for hand-held contact technique, but has never gained much popularity.

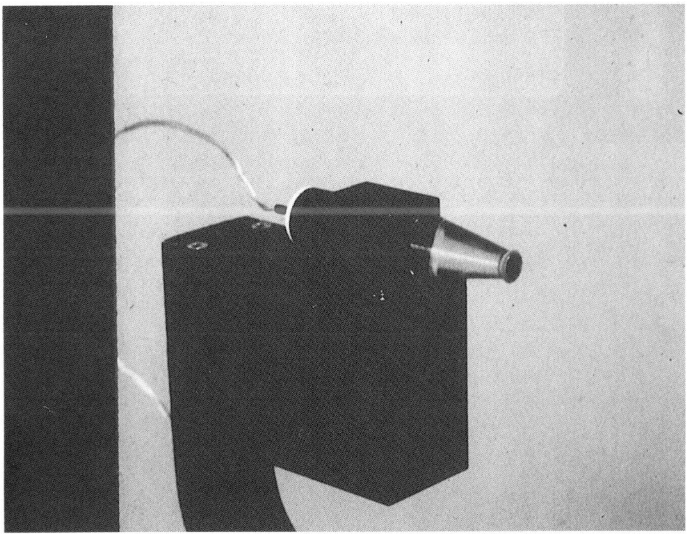

Figure 10: Water-filled, semi-soft probe that has been commonly used with the DBR unit.

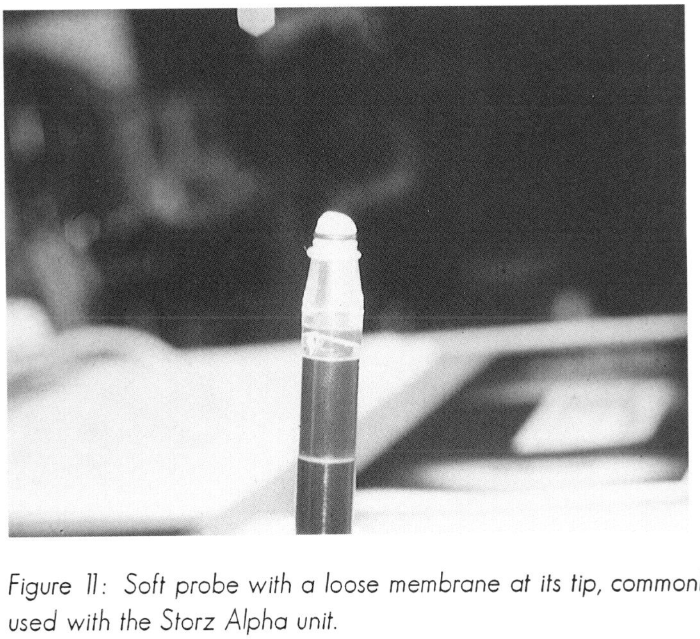

Figure 11: Soft probe with a loose membrane at its tip, commonly used with the Storz Alpha unit.

The transducer

The transducer consists of a quartz crystal located at the tip of the probe(7-9). The crystal has piezo-electric qualities that transform electric energy into mechanical energy in the form of sound waves and vice versa.

The crystal is connected by two electrodes to the main ultrasound system. Damping material is present behind the crystal to minimize reverberations of the ultrasound energy within the probe. Based on the piezo-electric principle, the transducer is an ultrasound source emitting a pulsed ultrasound beam. The performance of the crystal depends mainly on its shape, its diameter and its thickness.

Clinical Application:

(a) A transducer with a flat anterior surface will emit a non-focused beam (Fig. 12). This non-focused beam is an integral part of standardized diagnostic A-scan echography. The non-focused beam with its parallel borders allows accurate pattern recognition at all distances. During examination, the system sensitivity is decreased by 10 to 20 decibels; the beam width decreases substantially. The resultant narrow beam yields a more accurate axial length measurement.

(b) A transducer with a concave surface emits a focused ultrasound beam (Fig.13) essential for diagnostic B-scan echography because the examination of the vitreo-retinal layers are located within the beam's focal zone. Ultrasound manufacturers, using the popular B-scan technology, adopted the focused ultrasound beam within their biometry units. When a normal eye is measured, the retina is in the focal zone of the ultrasound beam yielding an accurate measurement. When a very short or a very long eye is measured, the retina is no more in the focal zone of the ultrasound beam causing difficulties in the pattern recognition and a less accurate measurement.

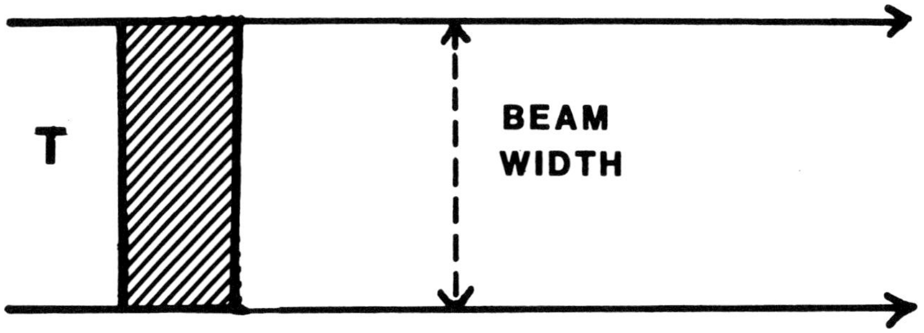

Figure 12: Non-focused beam emitted by a transducer (T) with a flat anterior surface.

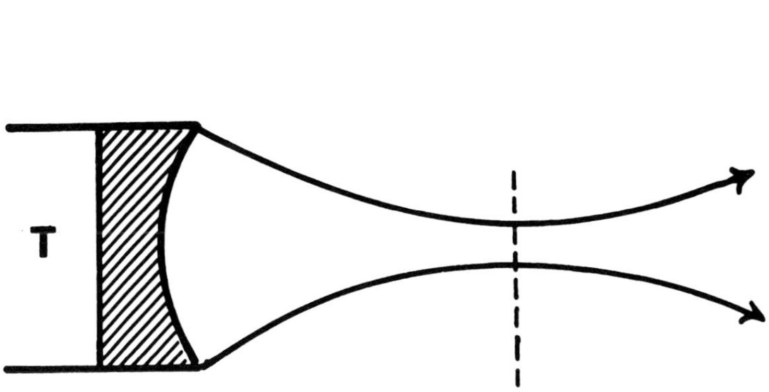

Figure 13: Focused beam emitted by a transducer (T) with a concave anterior surface.

The oscilloscope

The oscilloscope is the display screen that allows the technician to monitor the different echospikes generated by the ocular surfaces. Most units are equipped with automatic image-freezing capability when a valid measurement is detected and with gates for an electronic measurement of the axial length from the echogram displayed on the screen.

Clinical application:
Units without an oscilloscope are not recommended since they do not allow pattern recognition.

The sensitivity setting

The sensitivity setting controls the height of the echospikes displayed on the oscilloscope.

Clinical application:
The axial length is more accurately measured at a lower system sensitivity that allows a better pattern recognition of the anterior and posterior corneal surfaces, anterior and posterior lens surfaces, and anterior retinal surface (Fig. 14 A&B).

The velocity setting

The velocity setting controls the speed of sound propagation. The velocity, measured in meters per second (m/s) varies according to the medium through which sound propagates.

Clinical application:
Most units use an average velocity of 1548 m/s to 1556 m/s in a cataractous eye and 1532 m/s in an aphakic eye. Certain units measure each ocular compartment at its correct velocity: the lens thickness is measured with an average velocity of 1641 m/s while the anterior chamber and the viterous cavity's depth are measured with a velocity of 1532 m/s. These measurements are then computed within the instrument to display one axial length reading.

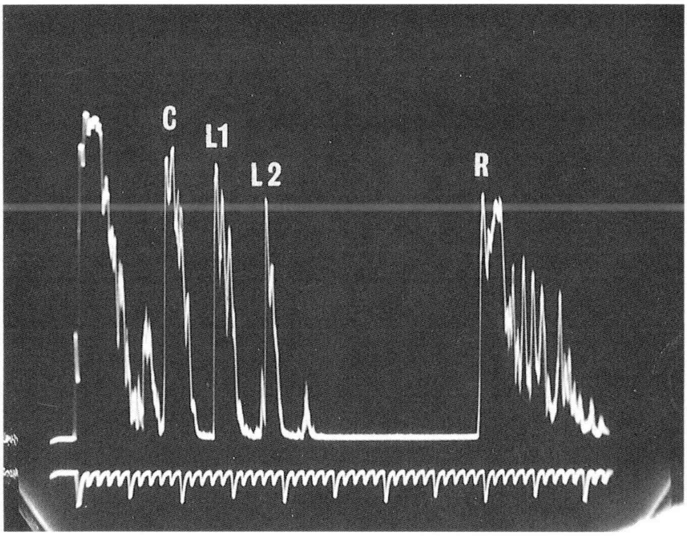

Figure 14 (A): Axial length measured at high system sensitivity.

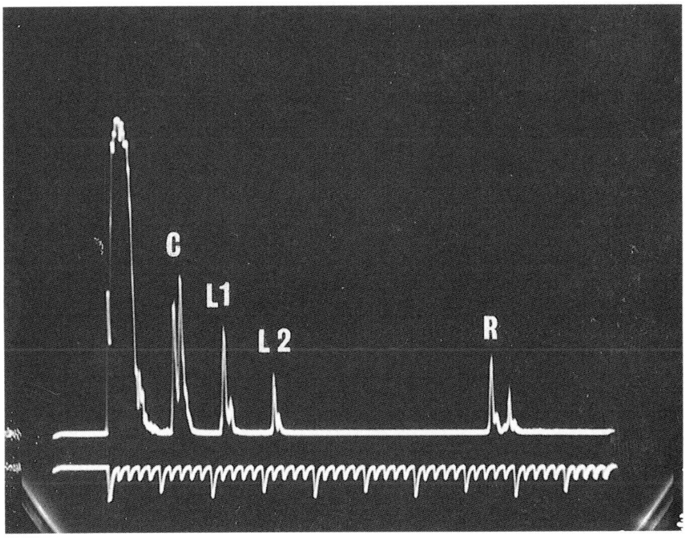

Figure 14 (B): Axial length measurement of the same eye at decreased system sensitivity for a better identification of the cornea (C), anterior (L1) and posterior (L2) lens surfaces, and of the anterior retinal surface (R).

The electronic gates

Electronic gates allow all newer ultrasound units to provide an electronic read-out of the axial length in millimeters. Biometers are equipped with two or four gates.

The two main gates are the "corneal gate" placed in the region of the corneal spike and the "retinal gate" placed in the region of the retinal spike (Fig. 15). Such instruments measure the travel time between the anterior surface of the cornea and the anterior surface of the retina and use an average sound velocity for the measurement of the axial length.

Other instruments are equipped with four gates which are usually thinner to allow a more accurate positioning of these gates over the leading edges (Fig.16A) or the peaks (Fig.16B) of the echoes generated from the anterior surface of the cornea, the anterior surface of the lens, posterior surface of the lens and anterior surface of the retina. Measurements of the anterior chamber depth, of the lens thickness and of the total axial length are displayed on the screen.

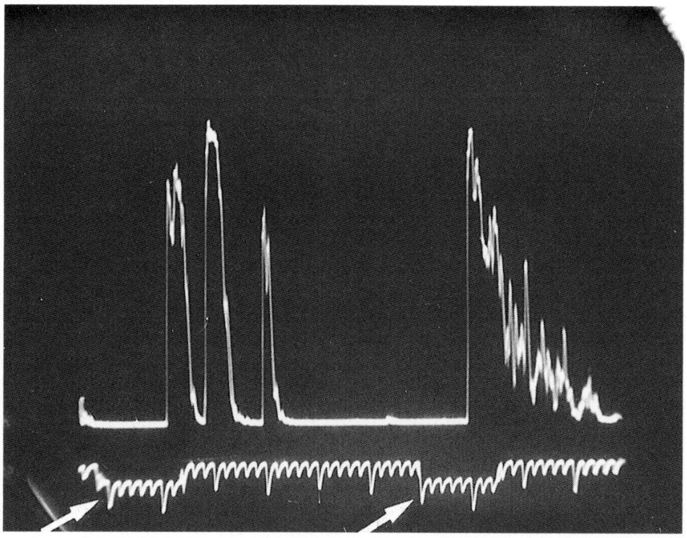

Figure 15: Indentation of the electronic scale (arrows) in the Kretz 7200 MA ultrasound unit equipped with a digital ocular computer. The first indentation represents the corneal gate and the second one, the retinal gate.

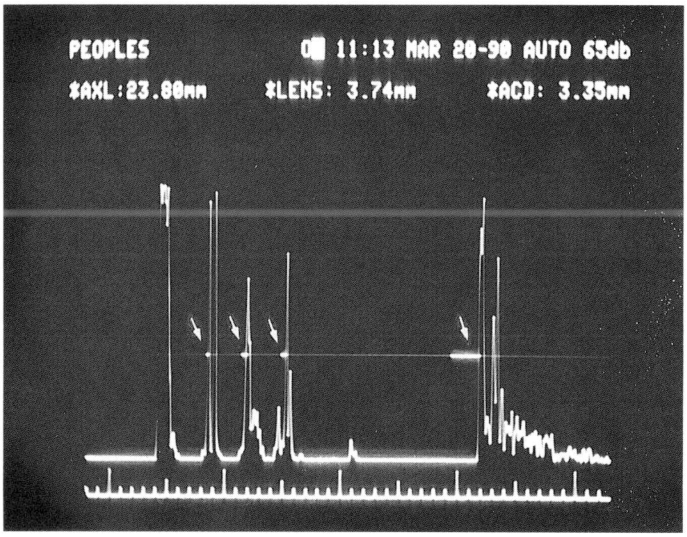

Figure 16 (A): Horizontal caliper lights used with an immersion technique, in the Innovation Imaging Inc. unit. The four gates (arrows) measure distances between the leading edges of the cornea, anterior lens surface, posterior lens surface and the retina.

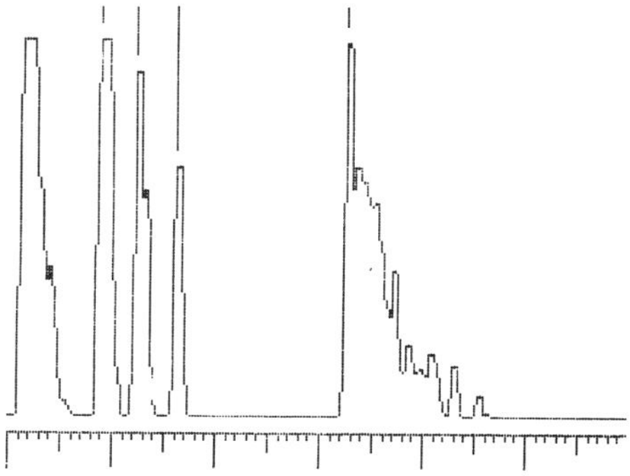

Figure 16 (B): Vertical caliper lines used with an immersion technique, in the Alcon Biophysic Ocuscan. The four gates measure distances between the ultrasound peaks of the cornea, anterior lens surface, posterior lens surface and the retina.

3. RETRIEVING AND REPORTING RESULTS

The A-scan echogram

When a measurement is taken, the A-scan echogram is displayed on the screen. This echogram is retrieved on a polaroid picture or a print-out from the instrument and kept on file. The advantages of such a practice are twofold:

-It allows the technician and/or the surgeon to review the echograms prior to surgery. If the measurement is not acceptable, it can then be repeated.

-It keeps a permanent record of the echogram, to be reviewed at a later date if the final result is not satisfactory.

Calculators and computer programs

Most formulas are programmed on calculators and computer diskettes to yield some or all of the following information:

- The IOL power for emmetropia
- The IOL power to produce a refraction of -1.0, -2.0 or -3.0D
- The IOL power to produce any amount of desired ametropia
- The IOL power for iseikonia
- The expected post-operative refractive error with any IOL
- The contact lens refraction if an implant is not used
- The IOL power with different formulas

These programs are also able to:

- Store information
- Average and compare results
- Review data
- Refine the anterior chamber constants.

Biometer's print-out

Newer biometers are equipped with built-in minicomputers that analyze the A-scan echograms and also perform IOL power calculations using one or more formulas programmed within the biometer. Once the calculations are done, a print-out is obtained for permanent record. These units notify the technician of some important values that, when present, should raise a red flag. The technician should then repeat the questionable measurement. The following values have been associated with a higher rate of errors in IOL power calculations:

- An axial length of less than 22 mm or over 25 mm, or a difference of over 0.3 mm between the two eyes.

- Average K readings of less than 40 D or over 47 D, or a difference of over 1 D between the two eyes.

- IOL power for emmetropia over 3 D from the anticipated power, or a difference of over 1 D between the two eyes.

4. CHOOSING THE APPROPRIATE BIOMETER

Most biometers provide reproducible and accurate measurements and are programmed with popular formulas. However, each biometer is characterized by specific features and components that have been discussed in length in this chapter. Here are some guidelines to assist in the choice of the appropriate biometer.

Five major features are of primordial importance:

- The biometer should have an oscilloscope or a display screen that allows pattern recognition of the displayed echogram and ensures an accurate and reproducible measurement.

- The biometer should allow measurement of the axial length in phakic, aphakic and pseudophakic eyes by setting the appropriate velocities.

- The biometer should provide a print-out or a picture of the A-scan echogram for review and safekeeping.

- The biometer should have clearly visible and easy-to-use electronic gates for a fast measurement of the axial length.

- The biometer should have a good service warranty. Most reputable manufacturers will provide a replacement unit while the original unit is being serviced.

In addition to these five important features, an ideal biometer will be equipped with a compact solid probe, fully mobile electronic gates, and will give the technician the choice of immersion or contact measurement techniques.

The following features should be looked at with a critical eye to decide their relative importance to the practice:

-The presence of freeze-frame capability. The A-scan echograms are frozen on the screen for easier recognition and measurement.

-The capability of measuring different ocular components such as the anterior chamber depth and the lens thickness.

-The presence of a built-in, user-friendly software programmed with the favorite formula (or formulas). It usually allows data storage and provides an easy-to-read print-out of the calculations. Although this convenient feature will increase the purchase price, it might improve efficiency.

-The presence of a diagnostic B-scan and/or pachymeter within the same chassis. Practices that handle vitreoretinal pathology and/or perform corneal surgery will enjoy the added features in one compact unit.

REFERENCES - CHAPTER IV

1. Shammas, HJ: Atlas of Ophthalmic Ultrasonography and Biometry. pp. 273-308. The C.V. Mosby Co., St. Louis, 1984.

2. Herrman, G and Buschmann, W: Methods of measuring the HF oscillation frequency in ultrasound pulses of equipment for diagnostic ultrasonography. Ophthalmol. Res. 3:274-282, 1972.

3. Till, P: Solid tissue model for the standardization of the echo-ophthalmograph 7200 MA (Kretztechnik). Doc. Ophthalmol. 41:205-240, 1976.

4. Ludwig, GD: The velocity of sound through tissues and the acoustic impedence of tissue. J. Acoust. Soc. Am. 22:862-866, 1950.

5. Lizzi, F, Burt, W and Coleman, DJ: Effects of ocular structures on the propagation of ultrasound in the eye, Arch. Ophthalmol. 84:635-640, 1970.

6. Lowe, R: Time amplitude ultrasonography for ocular biometry. Am. J. Ophthalmol. 66:913-918, 1968

7. Buschman, W: Special transducer probes for diagnostic ultrasonography of the eyeball. In Ultrasonics in ophthalmology, R. Goldberg and L. Sarin (eds.), Philadelphia, 1967, W.B. Saunders Co., pp. 87-101.

8. Coleman, DJ and Carlin, B: Transducer alignment and electronic measurement of visual axis dimensions in the human eye using time amplitude ultrasound. In Ultrasonics in ophthalmology, A.Oksala and H. Gernet (eds.), pp. 207-213, Basel, Switzerland,1967. S. Karger.

9. Gordon, D: Transducer design for ultrasonic ophthalmology. In Ophthalmic ultrasound, K. Gitter et al (eds.), pp. 65-66, St. Louis, 1969, the C.V. Mosby Co.

CHAPTER V

Axial Length Measurement
Patient Examination

The axial length (AL) is conventionally measured with ultrasonography, using a biometry unit. An ultrasound beam is emitted from the transducer located at the tip of the probe and whether an immersion or a contact technique is used, an A-scan pattern is displayed on the screen. A measurement is then taken.

1. EXAMINATION TECHNIQUES

Measurement of the axial length can be achieved with an immersion or a contact technique and, no matter which technique is used, it is strongly advisable to have both eyes measured for comparison purposes.

Immersion technique

The technique described herein is used with the Kretz 7200 MA unit(1-4). This technique also can be used with other ultrasound units equipped with a solid A-scan probe and mobile"gates" on the biometer's screen.

The patient is placed in a supine position on a flat examination table or in a reclining examination chair. A drop of local anesthetic is instilled in both eyes.

The proper scleral shell is chosen. These thin plastic cylinders are available in 16, 18, 20, 22 and 24 mm. diameter (Fig. 17). They can be purchased from Hansen Opticals (P.O. Box 613, Iowa City, IA 52244). While the 20 mm. shell fits most eyes, the larger cup provides a better fit in bigger eyes with large palpebral fissures and the smaller cup fits better in the presence of a narrow palpebral fissure. Other types of scleral shells are also available from different manufacturers. The flared edges of the scleral shell are placed between the lids, making sure that the cup is stable on the eye (Fig. 18).

Figure 17: The Ossoinig scleral shells used for the immersion technique.

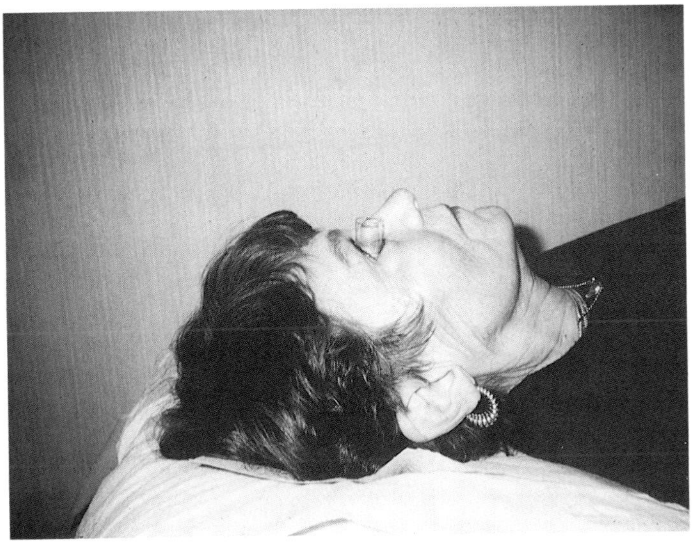

Figure 18: The patient is laying flat on her back and the flared edges of the cup are placed between the lids.

The cup is filled with gonioscopic solution. Methylcellulose 1% is preferred over the 2.5% concentration (too thick) and over saline solutions (too fluidy). The solution should be free of air bubbles; the presence of bubbles causes variations in the speed of sound and is responsible for noise formation within the ultrasound pattern. The easiest way to avoid bubbles is to remove the bottle's nipple and to pour the solution in the cup (Fig. 19). If bubbles do form within the solution, they are removed with a syringe (Fig. 20), and, if unsuccessful, the cup has to be emptied, cleaned, repositioned and refilled with gonioscopic solution.

The ultrasound probe is immersed in the solution keeping it 5 to 10 mm. away from the cornea (Fig. 21). The patient is asked to look, with the fellow eye, at a fixation point placed at the ceiling. Attention is then focused on the screen. The probe is gently moved until it is properly aligned with the optical axis and an acceptable A-scan echogram is displayed on the screen. A Polaroid picture or a print-out is obtained.

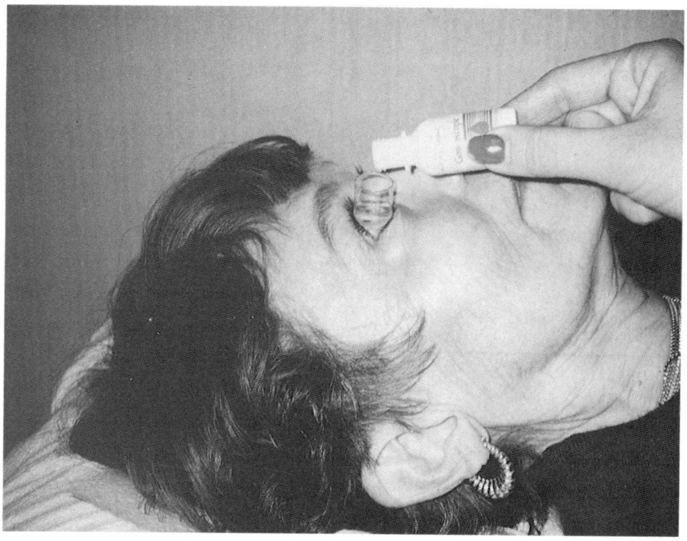

Figure 19: Methylcellulose 1% is poured into the cup.

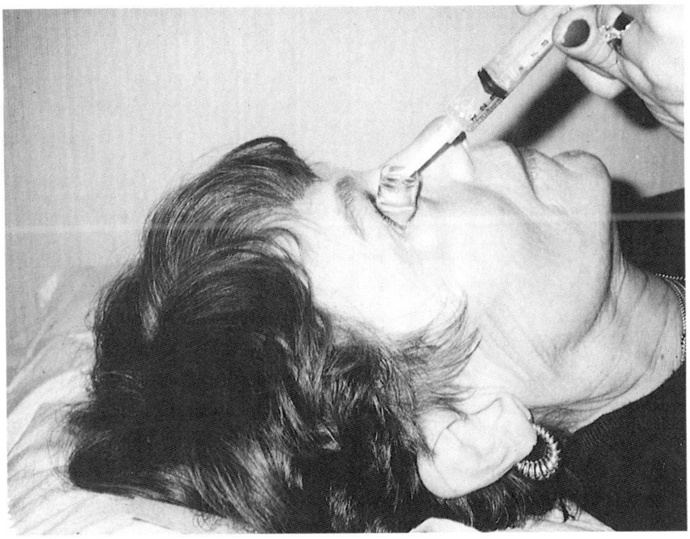

Figure 20: A syringe is used to remove any air bubbles. The same technique can be used to remove the gonioscopic solution at the end of the examination.

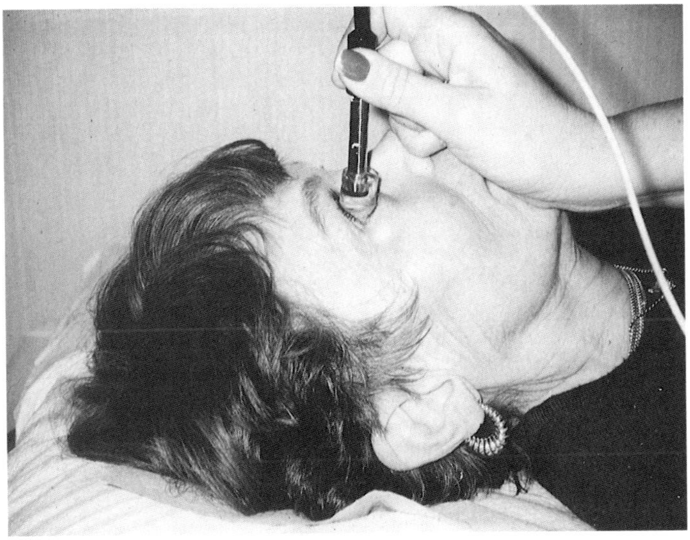

Figure 21: The solid probe is immersed in the solution keeping it 5 to 10mm away from the cornea.

Contact technique

A drop of local anesthetic is instilled in both eyes. The patient is examined in the seated position, with the chin correctly positioned on a free-standing chin rest. (5-7).

The ultrasound probe is attached to a zero-weight balance glide to prevent any pressure on the eye during examination. The probe's movements are controlled through a joystick handle (Fig. 22). The probe is positioned in front of the eye and the patient is asked to fixate on the red light within the probe. The probe is then brought forward to gently touch the cornea without indenting it (Fig. 23).

Attention is then focused on the screen. The probe is moved slightly up and down or to the side to optimize the echospikes displayed on the oscilloscope. A Polaroid picture or a print-out is obtained.

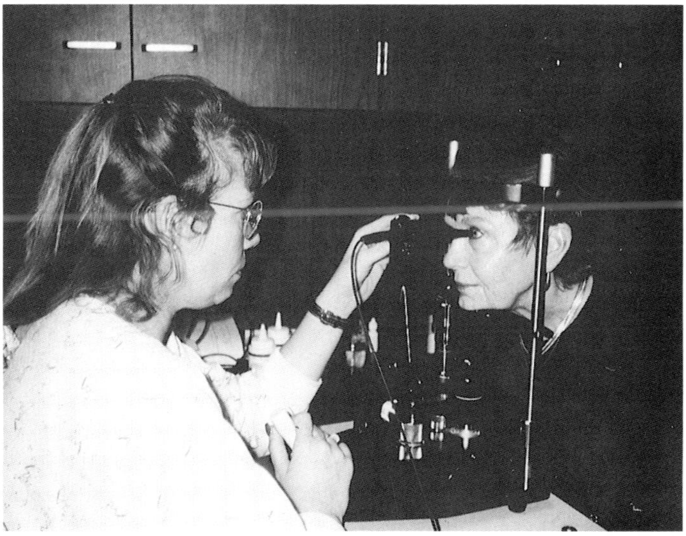

Figure 22: Technician uses the joystick to align the probe with the patient's eye.

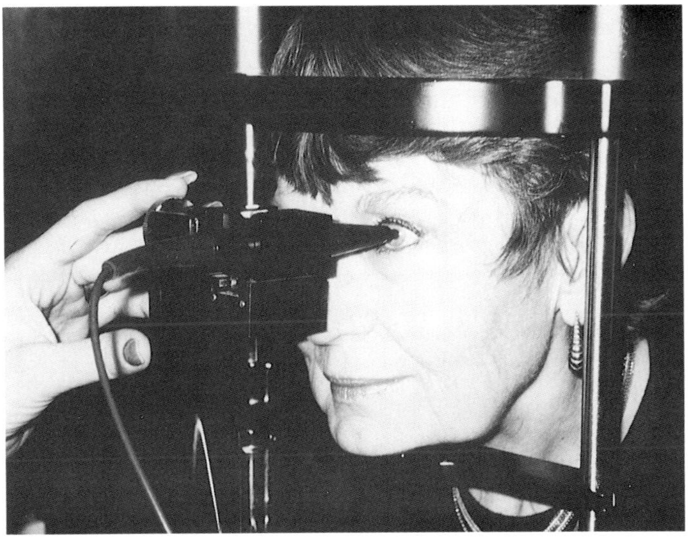

Figure 23: Contact axial length measurement using the solid probe.

Other probes are mounted on a Goldmann tonometer holder and are designed to be used with a slit-lamp (Fig. 24). The examination is conducted in the same manner as previously described.

The technique is slightly different for hand-held instruments. We herein describe our technique with the Bio-Pen(8), a hand-held, battery operated, compact biometric ruler (Fig. 25):

The patient is seated comfortably in an examining chair and a drop of local anesthetic is instilled in each eye. As the patient looks straight ahead, the Bio-Pen is activated by momentarily depressing a button. The examiner's hand is steadied against the patient's cheek and the Bio-Pen touches the cornea; a beep indicates contact. The ticking rate increases while the ultrasonic transceiver attempts to locate the four suprathreshold echoes from the anterior lens surface, posterior lens surface, retina and sclera. The built-in software of the unit compares the echographic pattern to an internal model of the human eye. Once four stable echoes are received which meet criteria set forth within the instrument, four dashes appear on the liquid crystal display. The axial length is read in millimeters (Fig. 26). Probes used with the "contact technique" should be cleaned after each examination to prevent contamination.

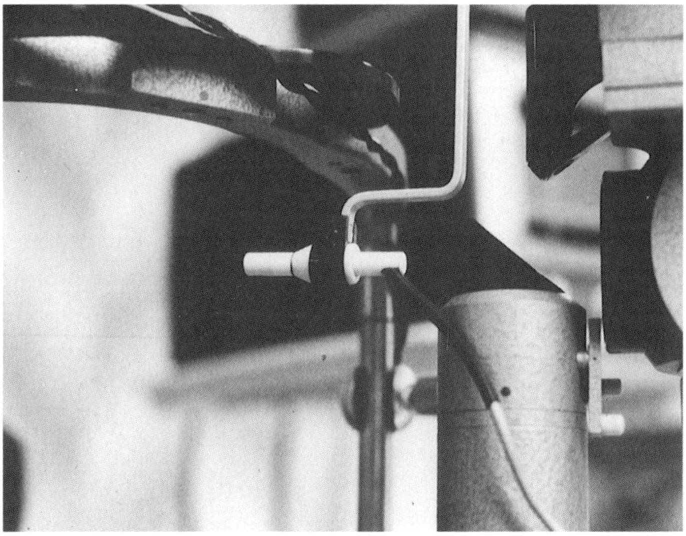

Figure 24: The small solid probe in the tonometer holder is used for axial length measurement.

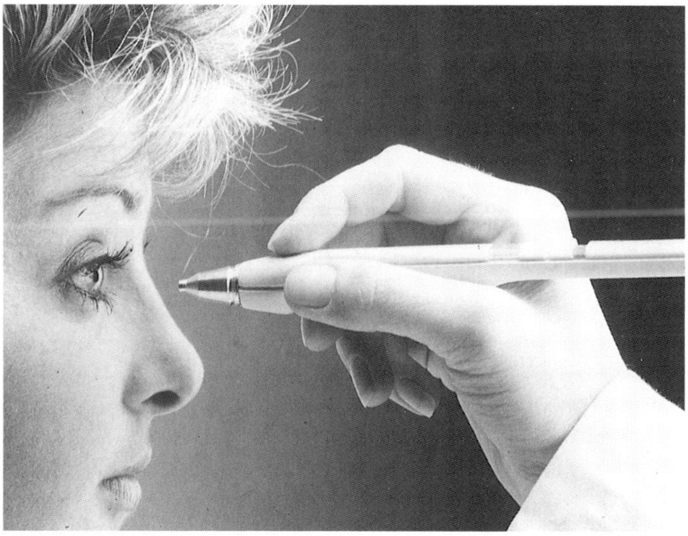

Figure 25: The hand-held Biopen is aligned with the visual axis of the eye.

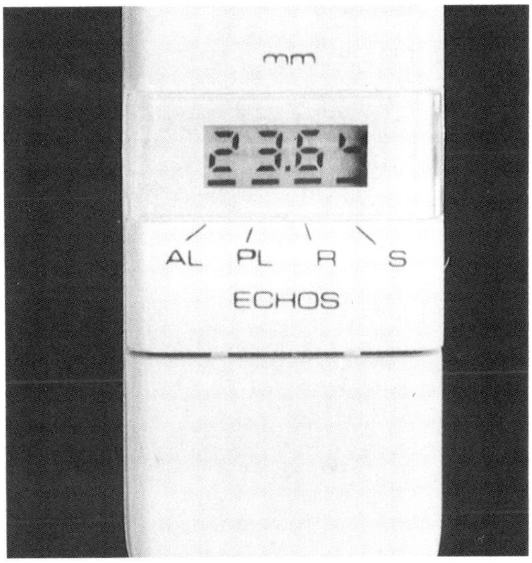

Figure 26: The axial length is read in millimeters. The four dashes below the reading represent four stable echospikes in the areas of the anterior lens surface, the posterior lens surface, the retina and the sclera.

Differences between examination techniques

The immersion and contact techniques for axial length measurement do not yield the same results. When measuring the same eye, the contact technique yields a shorter measurement than the immersion technique(9-11).

In a prospective study on 180 eyes (11), axial length measurement was obtained with both contact and immersion techniques. Each eye was measured with the Ocuscan-DBR (contact), the Ocuscan-400 (immersion), and the Kretz 7200 MA (immersion) units. Axial length measurements obtained with the contact technique were shorter than measurements obtained with the immersion technique by an average of 0.24 mm.

Figure 27 shows a scatter diagram comparing the shorter contact DBR measurements (y-axis) to the immersion-Kretz measurements (x-axis). All points are below the 45-degree angle equivalence line.

The differences between the two methods of examination include the patient's position and the possible corneal applanation by the ultrasound probe:

-The patient is conventionally examined in the seated position with the contact technique and the probe is brought forward to touch the cornea.

-The patient is conventionally examined in the supine position with the immersion technique and the solid probe is kept 5mm to 10 mm away from the cornea.

-These differences in the methods of examination, mainly the corneal indentation and the subsequent shallowing of the anterior chamber, are responsible for the shorter measurement obtained with the contact technique.

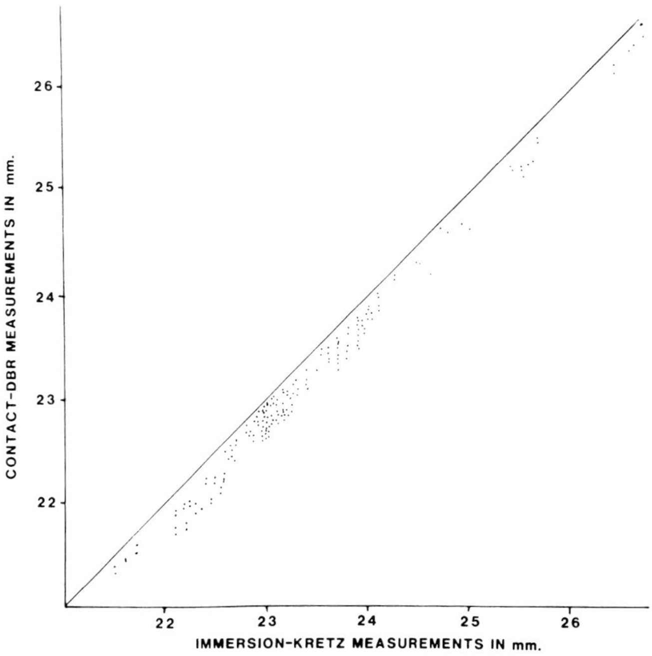

Figure 27: Scatter diagram in 180 axial length measurements comparing contact and immersion techniques. All measurements taken by the DBR contact unit were shorter than the ones taken with the Kretz-immersion unit. (Reprinted with permission from Shammas, HJ.: A comparison of immersion and contact techniques for axial length measurement. Am. Intraocular Implant Soc. J. 10:444-447, 1984.)

2. MEASUREMENT OF THE EYE

The A-scan pattern displayed during axial length measurement and the sound velocity used for this measurement vary whether the eye is phakic, aphakic, or pseudophakic.

The phakic eye

The A-scan pattern of a phakic eye(12-15) examined with an immersion technique displays the following echospikes from left to right (Fig. 28 A & B):

- IS: The initial spike (IS) represents reverberations at the tip of the probe with no clinical significance. Many units will allow the technician to move the whole A-scan pattern to the left and remove the IS from the picture (Fig. 28 B).
- C: The corneal spike (C) is double peaked representing the anterior and posterior surfaces of the cornea.
- L1: The anterior lens spike (L1) is generated from the anterior surface of the lens.
- L2: The posterior lens spike (L2) is generated from the posterior surface of the lens, and is usually smaller than L1.
- R: The retinal spike (R) is generated from the anterior surface of the retina. It is straight, high-reflective and tall whenever the ultrasound beam is perpendicular to the retina, as it should be during axial length measurement.
- S: The scleral spike (S) is another high-reflective spike generated from the scleral surface, right behind the retinal spike, and should not be confused with it.
- O: The orbital spikes (O) are low-reflective and located behind the scleral spike.

The A-scan pattern of a phakic eye examined with the contact technique demonstrates similar echospikes except that the corneal spike is merged with the initial spike since the cornea is in touch with the tip of the probe.

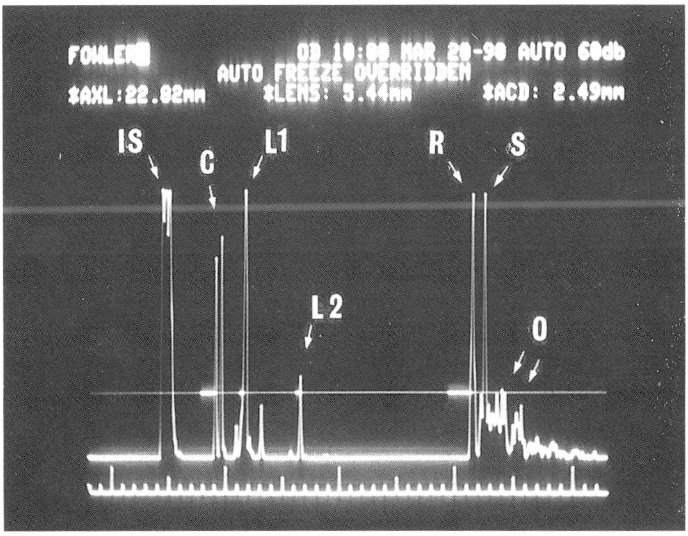

Figure 28 (A): *Ultrasound display of the different echospikes during axial length measurement, identifying the initial spike (IS) the cornea (C), the lens surfaces (L1 and L2), the retina (R), sclera (S) and orbital tissues (O).*

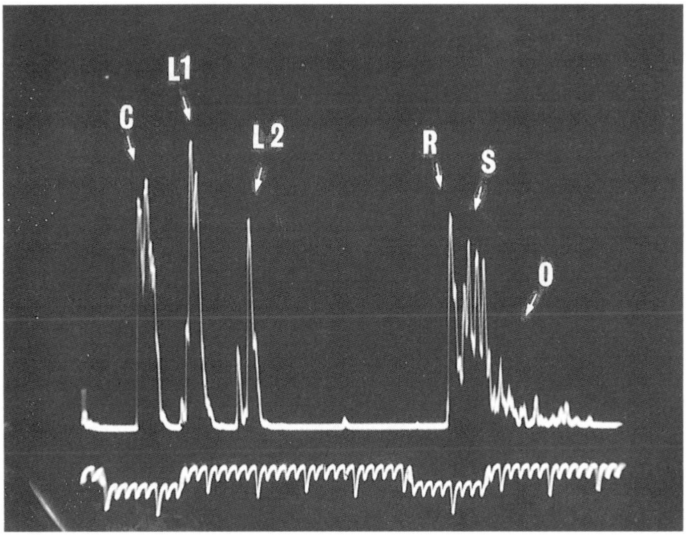

Figure 28 (B): *A-scan pattern of a phakic eye where the initial spike has been removed from the screen display.*

Some biometers give the readings directly in millimeters using an average sound velocity. This velocity is reported in meters per second(m/s). Most biometers use an average velocity of 1548 to 1556 m/s (16). This author uses a velocity of 1553 m/s that has been reached by reviewing the results of 1000 axial length measurements in cataractous eyes(17).

Many modern biometers use separate sound velocities for the different eye components to obtain the total axial length(16). The eye is divided ultrasonically into four compartments (Fig.29):

-The corneal thickness(18) is measured between the anterior (Cl) and posterior (C2) surfaces of the cornea using a velocity of 1620 m/s. The thickness of a normal cornea is approximately 0.5 mm.

-The true anterior chamber depth(19) is measured between the posterior corneal surface (C2) and the anterior lens surface (L1) using a velocity of 1532 m/s. This measurement is not to be confused with the anterior chamber constant used in the IOL power calculation formulas which usually represents the sum of the corneal thickness and of the true anterior chamber depth.

-The lens thickness(20-25) is measured between the anterior lens surface (L1) and the posterior lens surface (L2) using a velocity of 1641 m/s. Actually 1640.5 m/s is the calculated sound velocity in the normal crystalline lens. The sound velocity varies in cataractous eyes with a slower velocity (average 1590 m/s) in the intumescent cataracts due to their high water content, and a higher velocity in the posterior capsular cataracts. In most cases of nuclear sclerosis with subcapsular changes, the sound velocity averages 1641 m/s.

-The vitreous cavity's depth(19) is measured between the posterior lens surface (L2) and the anterior surface of the retina (R) using a velocity of 1532 m/s.

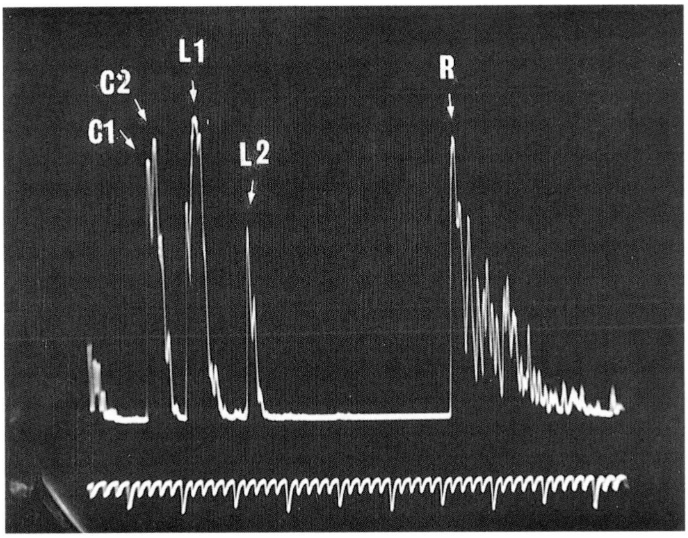

Figure 29: A-scan pattern of a phakic eye, taken with the Kertz 7200 MA unit, and with the Initial Spike removed from the screen display. The two corneal peaks are identified as C1 and C2, the anterior lens spike as L1, the posterior lens spike as L2 and the retinal spike as R.

A manual measurement of the axial length(16) is used with older ultrasound units not equipped with an electronic read-out, and is rarely used nowadays. Using calipers, a measurement is taken from the scale in microseconds. However, this reading represents the travel time it takes the ultrasound beam to reach the tissue under examination and return to the probe, thus representing twice the actual measurement. It has then to be divided by 2, then converted to millimeters using an average velocity.

The aphakic eye

In an aphakic eye(15,26), the A-scan pattern displays the following echospikes from left to right (Fig. 30A):

IS: The initial spike. It can be moved out of the screen in an immersion technique and it will be merged with the corneal spike (C) in a contact technique.
C: The double-peaked corneal spike
I: A medium reflective echospike from the iris surface and/or anterior vitreous face.
R: The retinal spike from the anterior surface of the retina.
S: The scleral spike.
O: The orbital spikes.

The initial spike can also be removed from the screen (Figure 30B) leaving only the echospikes generated from the eye.

The axial length is measured between the anterior corneal surface (C) and the anterior retinal surface (R) using an average sound velocity of 1.532 mm/mis (1532 m/s), which is the velocity in aqueous and vitreous. Certain units use a slightly higher sound velocity of 1534 m/s to account for the faster speed of sound within the cornea.

If the ultrasound unit uses only fixed 1550 m/s velocity and does not allow for the use of a 1534 m/s velocity, the axial length of the aphakic eye can then be calculated:

Aphakic AL = (1534/1550) X AL measured with 1550 m/s

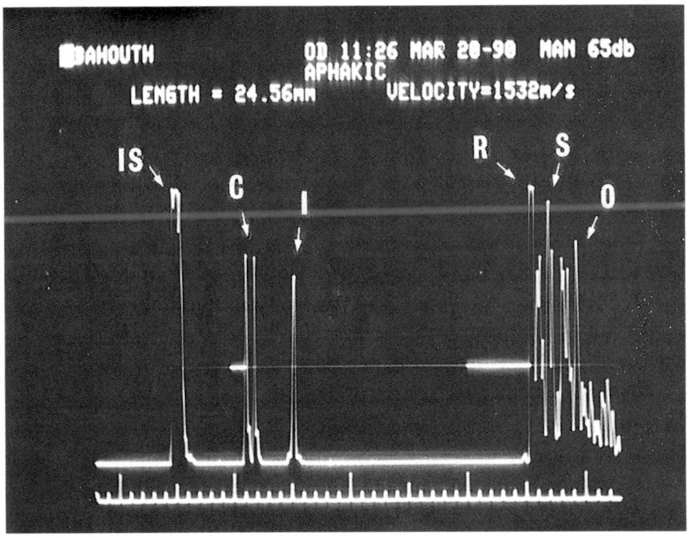

Figure 30 (A): A-scan pattern of an aphakic eye, identifying the initial spike (IS), the cornea (C), the iris and/or anterior vitreous face (I), the retina (R), sclera (S) and orbital tissues (O).

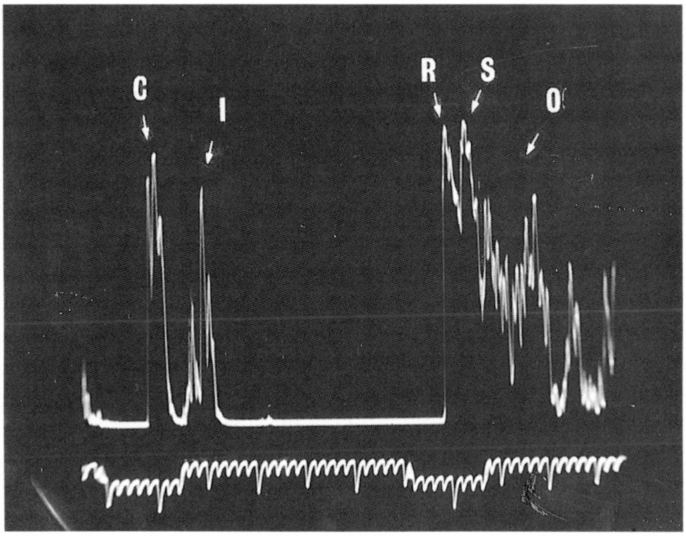

Figure 30 (B): A-scan pattern of an aphakic eye where the Initial Spike has been removed from the screen display.

The pseudophakic eye

In a pseudophakic eye(26-29), the A-scan pattern displays the following echospikes from left to right (Fig. 31):

IS: The initial spike. It can be moved out of the screen in an immersion technique. It is merged with the corneal spike (C) in a contact technique.
C: The double-peaked corneal spike
P: A high reflective spike from the anterior surface of the pseudophakic lens. It is usually followed by multiple smaller echospikes (M) that represent reverberations of the ultrasound beam between the anterior and posterior surfaces of the implant.
R: The high reflective retinal spike
S: The scleral spike
O: The orbital spikes

The axial length of a pseudophakic eye(28-30) is calculated:

$$AL = MAL\ 1532 + T\ (1 - 1532/Vel)$$

Where: MAL 1532 is the axial length measurement taken at a velocity of 1532 m/s, T is the central thickness of the pseudophakic lens and Vel the sound velocity within the pseudophakic lens.

If the measurement of the axial length is taken at a velocity of 1550 m/s (MAL 1550), it can easily be converted:

$$MAL\ 1532 = (1532/1550) \times MAL\ 1550$$

The average sound velocities (Vel) and central thickness (T) of the different implants are as follows:

Implant	Sound Velocity	Central Thickness
PMMA	2660 m/s	0.6 to 0.8 mm
Silicone	980 m/s	1.2 to 1.5 mm
Glass	6040 m/s	0.3 to 0.4 mm
Acrylic	2200 m/s	0.7 to 0.9 mm

If the eye is to be measured with an average sound velocity instead of using the preceding formula, the following values are recommended:

- 1555 m/s for an eye with a PMMA IOL
- 1476 m/s for and eye with a silicone IOL
- 1549 m/s for an eye with a glass IOL
- 1554 m/s for an eye with an acrylic IOL

If a pseudophakic eye is measured at the phakic average velocity of 1553 m/s, the error is less than 0.1 mm for the eye with a PMMA, glass or acrylic IOL. However, this error exceeds 1.0 mm for the eye with a silicone IOL(30, 31).

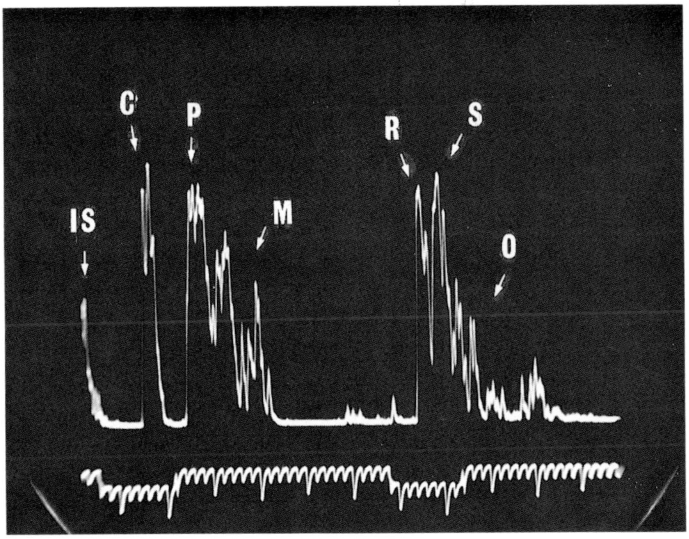

Figure 31: A-Scan pattern of the pseudophakic eye identifying the initial spike (IS), the cornea (C), the pseudophakos lens (P), the multiple reverberations (M), the retina (R), sclera (S) and orbital tissues (O).

3. VARIATIONS IN AXIAL LENGTH MEASUREMENT

Variations in axial length measurement are due to the use of different examination techniques and to the use of different sound velocities by the biometer.

Variations due to the examination technique

The immersion and contact techniques for axial length measurement are described in this chapter. When measuring the same eye, the contact technique yields a shorter measurement than the immersion technique. The difference varies between 0.1 mm and 0.3 mm depending on the probe used with the contact technique, and how much pressure is applied to the eye during examination(8,11).

Variations due to changes in the sound velocity

Although it is best to measure the different ocular compartments at their specific sound velocities, the use of an average sound velocity of 1553 m/s yields clinically insignificant errors in the measurement.

In the average 23.5 mm eye, the use of an average 1553 m/s sound velocity will yield the exact same measurement as the use of specific sound velocities for the aqueous, lens and vitreous; this is true if the anterior chamber depth and lens thickness have average values. The use of a lower (1548 m/s) or higher (1556 m/s) sound velocity can vary the axial length by less than 0.05 mm.

In the long 26.0 mm eye with an anterior chamber of 3.2 mm and a lens thickness of 4.2 mm, the use of the average 1553 m/s yields a measurement of 26.04 mm, with an error of only 0.04 mm.

In the short 21.0 mm eye with an anterior chamber of 2.5 mm and a lens thickness of 4.8 mm, the use of the average 1553 m/s yields 20.93 mm, with an error of only 0.07 mm.

Approximation of an axial length

Occasionally, the eye to be operated on is difficult to measure, whether it is a recently traumatized eye or an eye with vitreoretinal pathology where the retina can not be identified. The axial length can be approximated:

- -The opposite eye is measured.
- -The measurement is adjusted for the difference in refraction provided, of course, that there is no difference in the corneal curvature. The original refraction must be used for accuracy.
- -Add 0.1 mm to the axial length for each -0.25 D. of additional myopia or subtract 0.1 mm for each +0.25 D. of additional hyperopia.
- - If the eye has a scleral buckle for treatment of a detached retina, add 0.3 mm to the axial length.

In most cases where the axial length cannot be measured and has to be approximated, the eye is compromised and the chances of visual recovery to the 20/40 level are quite poor.

Measurement of an eye with an intumescent cataract

In the presence of an intumescent cataract(24,25), the lens increases its water content and becomes thicker (over 5.0 mm). Concomitantly, the sound velocity within the lens decreases to around 1590 m/s from the usual 1641 m/s.

When the axial length is measured using separate sound velocities, the erroneous use of a 1641 m/s sound velocity will yield a 0.15 mm longer measurement, calling for a weaker IOL and resulting in a +0.40 to +0.50 more hyperopic final refraction. In these cases it is better to use the average velocity. The decrease in the sound velocity within the lens and the increase in the lens thickness leave the average sound velocity between 1549 m/s and 1552 m/s which is very similar to the average velocity within an eye with nuclear sclerosis.

Measurement of an eye with silicone-filled vitreous

In an eye where the vitreous is filled with liquid silicone(30), the retinal echospike is small and difficult to display due to sound attenuation within the liquid silicone. The system sensitivity should be increased to better identify the retinal spike.

The measurement of such an eye is more difficult. The sound velocity in liquid silicone varies between 970 and 1100 m/s, depending on the type and amount of silicone used. An average value of 990 m/s is used for calculation purposes. It would be best to measure each component of the eye separately for a more accurate result. This will require a measurement with an instrument that allows you to do so, such as the Kretz unit, with its electronic scale.

An average velocity for an eye where the vitreous cavity is filled with liquid silicone has been calculated to be 1133 m/s. If the usual average velocity of 1553 m/s is used instead of the calculated 1133 m/s, the 23.5 mm eye will be measured as 33.2 mm with an error of over 8 mm.

Many ultrasound units measure the anterior chamber depth and lens thickness at their proper velocities and also give a display of the total axial length. The vitreous cavity's depth (VCD) (measured with 1532 m/s) is calculated by subtracting the anterior chamber depth and the lens thickness from the measured axial length (AL). The true measurement of the vitreous cavity's depth is then calculated for a velocity of 990 m/s:

$$\text{True VCD} = 990/1532 \times \text{VCD measured with } 1532\ \text{m/s}$$

The true vitreous cavity's depth is then added to the lens thickness and anterior chamber depth to derive the true axial length. However, this method can only be used if silicone is filling the totality of the vitreous cavity. If silicone is only partially filling the vitreous cavity, then separate measurements are needed.

Effects of axial length variations on the final refraction

These variations in axial length measurement affect the final refraction differently in the average, long and short eyes(15).

-In an average 23.5 mm eye, a 0.1 mm difference in axial length measurement affects the final post-operative refraction by 0.25 D. When using a contact technique, the eye is measured shorter by around 0.24 mm, calling for the use of a 0.76 D. stronger IOL and resulting in a -0.60 D. more myopia in the final refraction than expected. However all IOL formulas allow for an adjustment in the A constant (SRK formulas), in the S factor (Holladay's formula) or in the anterior chamber (AC) depth (most theoretical formulas) to adjust for this consistent difference in the A.L. measurement.

-In a long 26.0 mm eye, a 0.1 mm difference in the A.L. measurement affects the final post-operative refraction by only 0.20 D.

-In the short 21.0 mm eye, a 0.1 mm difference in the A.L. measurement affects the final post-operative refraction by 0.31 D.

Although the immersion and contact techniques yield consistent and reproducible measurements, the routine measurement of the axial length with an immersion technique has been very rewarding, especially in the short and long eyes.

REFERENCES - CHAPTER V

1. Byrne, SF: Standardized echography, Part I: A-Scan examination procedures, Int. Ophthalmol. Clin. 19:267-281, 1979.

2. Ossoinig, KC: Standardized echography: basic principles, clinical applications and results Int. Ophthal. Clin. 19:127-210, 1979.

3. Shammas, HJF: Axial length measurement and its relation to intraocular lens power calculations. Am. Intraoc. Implant Soc. J.8:346- 349, 1982.

4. Kendall, CJ: Ophthalmic Echography (The Ophthalmic Technical Series), pp. 57-106, Thorofare, NJ, 1990.

5. Binkhorst, RD: Biometric A-Scan ultrasonography and intraocular lens power calculation, in Current Concepts In Cataract Surgery: selected proceedings of the Fifth Biennial Cataract Surgical Congress, pp. 175-182, 1987 Mosby CV.

6. Coleman DJ: Ophthalmic biometry using ultrasound, Int. Ophthalmol. Clin. 9:667-683, 1969.

7. Coleman, DJ and Carlin, B: A new system for visual axis measurements in the human eye using ultrasound, Arch. Ophthalmol. 77:124-127, 1967.

8. Shammas, HJ, and Swearingen, M: Clinical evaluation of the Bio-Pen for axial length measurement. J. Cat. Refr. Surg., 16:120-122, 1990.

9. Olsen, T., Nielsen PJ: Immersion versus contact technique in the measurement of axial length by ultrasound. Acta Ophthalmologica, 67:101-102, 1989.

10. Schelenz,J, Kammann, J: Comparison of contact and immersion techniques for axial length measurement and implant power calculation. J. Cataract and Refractive Surgery, 15:425-428, 1989.

11. Shammas, HJF: A comparison of immersion and contact techniques for axial length measurement. American Intraocular Implant Society Journal, 10:444-447, 1984.

12. Purnell, EW and Sokollu, A: Ultrasonic measurements of eye length. Acta Ophthalmol. 40:219-222, 1962.

13. Gernet, H: Biometry of the eye by means of ultrasound, Klin. Monatsbl. Augenheilkd. 146:863-869, 1965.

14. Gernet, H:, and Franceschetti, A: Ultrasound biometry of the eye (review). In Ultrasonics in ophthalmology, A. Oksala and H.Gernet (eds.), pp. 175-200, Basel, Switzerland,1967, S. Karger

15. Shammas, HJ: Atlas of Ophthalmic Ultrasonography and Biometry. pp. 273-308. The C.V. Mosby Co., St. Louis, 1984.

16. Shammas, HJF: Manual versus electronic measurement of the axial length. In Ultrasonography in Ophthalmology, pp. 225-229, Proceedings of the 1982 Ninth SIDUO Congress J.S. Hillman and M.M. LeMay, Editors. The Hague,1983. Dr. W. Junk Publishers.

17. Shammas, HJF: A-Scan biometry of 1000 cataractous eyes. In "Ophthalmic Echography". Proceedings of the 10th SIDUO Congress. K.C.

Ossoining, Editor. Dr. W. Junk, Publisher. Doc. Ophthalmologica 48:57-63, 1987.

18. Delmarcelle, Y, Colignon-Brach, J and Luyckx-Bacus, J: The role of the cornea and lens in biometry of the anterior chamber in normal subjects, Arch. Ophthalmol. 30:291-300, 1970..

19. Bell, DS: Approximate determination of the speed of sound through liquids and solids, J. Acoust. Soc. Am. 44:656,1968.

20. Oksala, A, and Lehtinen, A: Measurement of the velocity of sound in some parts of the eye, Acta Ophthalmol, 36:633-639, 1958.

21. Jansson, F, and Kock, E: Determination of the velocity of ultra sound in the human lens and vitreous, Acta Ophthalmol. 40:420-433, 1962.

22. Coleman, DJ, Lizzi, FL, Franzen, LA and Abramson, DH: A determination of the velocity of ultrasound in cataractous lenses, Bibl. Ophthalmol. 83:246-251, 1975.

23. Laursen, AB, and Fledelius, H: Variations of lens thickness in relation to biomicroscopic types of human senile cataract, Acta Ophthalmol. (Copenhagen) 57:1-13, 1979.

24. Pallikaris, I, Gruber, H: Determination of sound velocity in different forms of cataracts. Documenta Ophthalmologica 29:165-169,1981.

25. Massin, M, Lambrinakis, I: In vivo determination of the speed of ultrasound in cataractous lenses. In Ultrasonography in Ophthalmology, 12:131-134, 1990.

26. Olsen, T: Calculating axial length in the aphakic and the pseudophakic eye. J. Cataract and Refractive Surgery, 14:413-416, 1988.

27. Freudigor, H, Artaria, L, Niesel, P: Influence of intraocular lenses on ultrasound axial length measurement: In vitro and vivostudies. Am Intraocular Implant Soc. J., 10:29-34, 1984.

28. Holladay, JT, Prager, TC: Accurate ultrasonic biometry in pseudophakia. American J. Ophthalmology, 107:189-190, 1989.

29. Naeser, K, Naeser, A, Boberg-Ans, J, Bargum, R: Axial length following implantation of posterior chamber lenses. J. Cataract Refractive Surg. 15:673-675, 1989.

30. Hoffer, KJ: Ultrasound velocities for axial length measurement. J. Cat. Refr. Surg. 20:554-562, 1994.

31. Milauskas, AT, Marney, S: Pseudo axial length increase after silicone lens implantation as determined by ultrasonic scans. J. Cataract and Refractive Surgery, 14:400-402, 1988.

CHAPTER VI

AXIAL LENGTH MEASUREMENT: AVOIDING ERRORS

In an accurate axial length (AL) measurement, the technician aligns the ultrasound beam with the optical axis of the eye by being perpendicular to the four major surfaces of the eye: the anterior surface of the cornea, the anterior surface of the lens, the posterior surface of the lens and the anterior surface of the retina. Biometry units that are not equipped with an oscilloscope or a screen that displays the actual scan have a high error rate and are definitely not recommended.

Errors in an AL measurement are due to an improper technique yielding shorter or longer measurements(1-3). Often, manufacturers recommend using the average value of multiple measurements to avoid errors. Although this is a good practice, one should remember that multiple readings of an erroneous measurement will still yield an erroneous average measurement. Occasionally a patient can be uncooperative during biometry. The technician should use good judgement and use the best technique to achieve the task. The "shakers" and "squeezers" will present the technician with a challenge. A contact technique is sometimes almost impossible to perform and an inaccurate measurement is taken between two lid squeezes. An immersion technique with a scleral shell keeps the lids open for an easier measurement. However, the technician should remember that the two techniques for axial length measurement do not yield exactly the same results and an adjustment must be made accordingly in the IOL formula.

1. SHORTER AXIAL LENGTH MEASUREMENT

Shorter measurements of the axial length can occur with corneal compression, off-axis measurement, wrong placement of the electronic gates and/or the presence of vitreous pathology.

Entering shorter measurement of the AL in IOL power calculations will call for the use of a stronger IOL than is actually required, resulting in an induced myopia in the final post-operative refraction.

Corneal compression

Corneal compression(4,5) is the most common cause of shorter AL measurements. An unskilled technician can indent the cornea with the A-Scan probe more than needed, resulting in a shallower anterior chamber depth and a shorter axial length, even though an acceptable A-Scan echogram has been displayed the screen.

Corneal compression is more pronounced with hand-held probes where there is no chin rest to stabilize the patient's head, and with large, flat surfaced probes that have the tendency to applanate a larger surface of the cornea. Corneal compression is less accentuated with the newer, smaller probes that have a concave surface to mold the anterior corneal surface.

Case presentation:

Figures 32 A&B are taken from the same eye. Although both scans look acceptable, note that the AL is 0.3mm shorter in figure 32A where the technician has compressed the cornea during the examination.

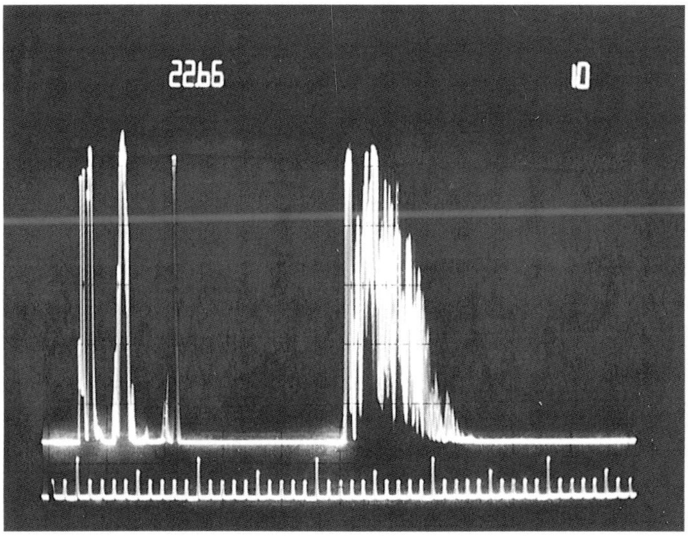

Figure 32 (A): A-scan display of a phakic eye with the corneal compression showing an axial length reading of 22.65 mm.

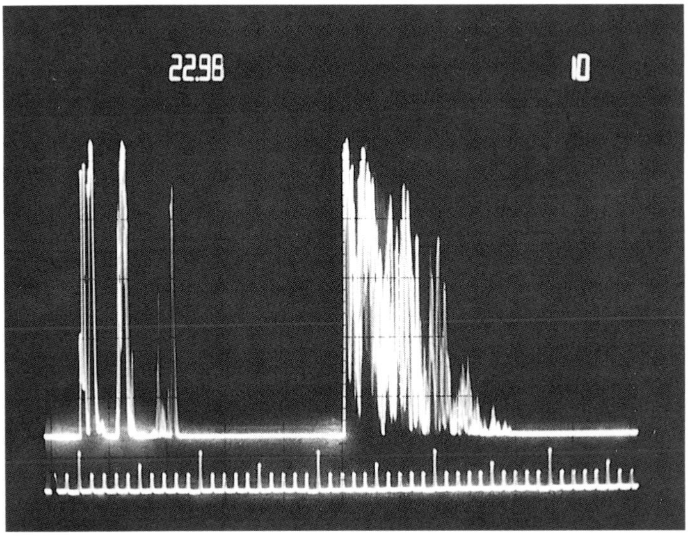

Figure 32 (B): A-scan display of the same eye with no corneal compression showing an axial length reading of 22.96 mm.

Off-axis measurement

Off-axis measurement(6,7) occurs when the ultrasound beam is not perpendicular to the surfaces of the eye.

(a) A minimal off-axis scan is characterized by the absence of the posterior lens spike or the presence of a very small one. The remainder echospikes from the cornea, anterior surface of the lens and from the retina usually appear normal.

Case presentation:
Figures 33 A and B are taken with the contact technique from the same eye. Note the small posterior lens spike in Figure 33A due to a slight off-axis ultrasound direction resulting in a 0.14mm shorter measurement when compared with the ultrasound pattern in Figure 33B.

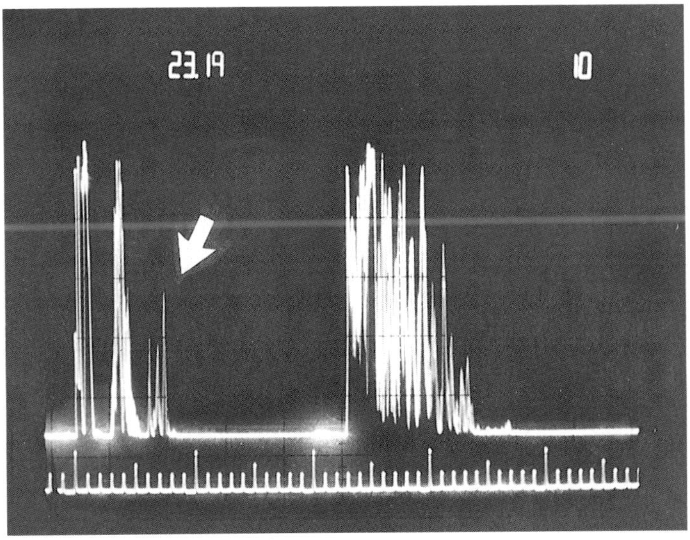

Figure 33 (A): A-scan display of a phakic eye with a slight off-axis measurement of 23.19mm. Note the presence of the small posterior lens surface echospike (arrow).

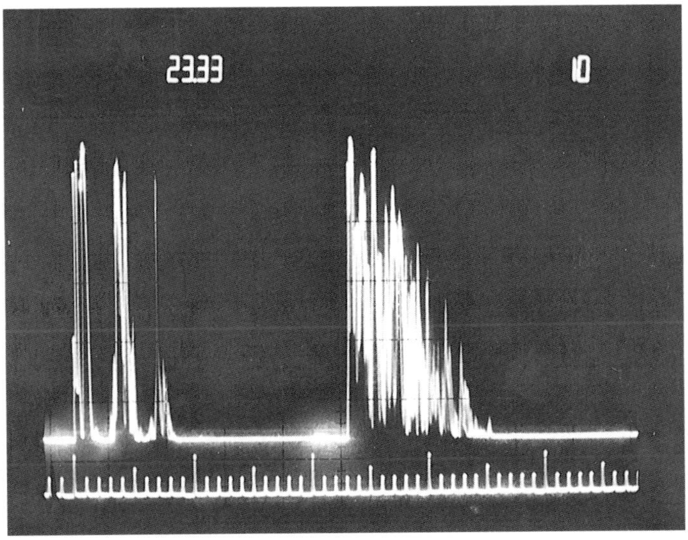

Figure 33 (B): A-scan display of the same eye with an on-axis measurement of 23.33mm.

(b) A larger off-axis measurement occurs when the patient is not looking at the fixation light due to the presence of a dense cataract or to the inability of the patient to hold the eye in a steady position. A larger off-axis scan is characterized by the absence of the posterior lens spike and the presence of a jagged retinal spike.

Case presentation:
Figures 34 A and B are taken with the contact technique from the same eye. Note the small lens spikes and the jagged retinal spike in Figure 34A due to a large off-axis ultrasound direction resulting in a 0.51mm shorter measurement when compared with the ultrasound pattern in Figure 34B.

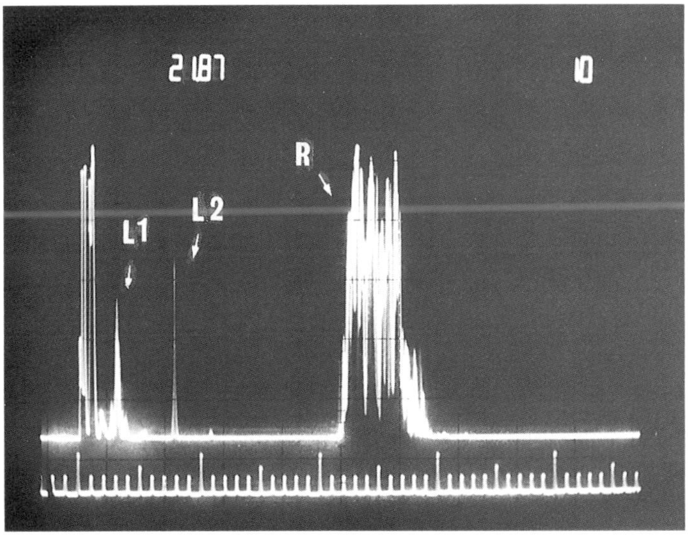

Figure 34 (A): A-scan display of a phakic eye with a large off-axis measurement of 21.87mm. Note the presence of the small lens spikes (L1 & L2) and the jagged retinal spike (R).

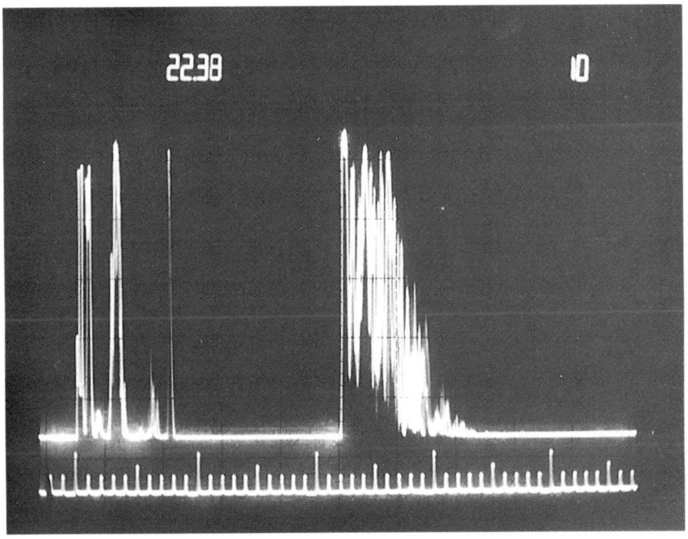

Figure 34 (B): A-scan display of the same eye with an on-axis measurement of 22.38mm.

(c) Off-axis scan with the immersion technique causes additional changes to the corneal spike.

Perpendicularity to the corneal surface can not be evaluated with the contact technique since the corneal spike is merged with the very strong initial spike originating at the tip of the probe. The corneal spike can only be evaluated with the immersion technique where the cornea remains 5 to 10 mm away from the tip of the probe. In a large off-axis measurement, the corneal spike is very small and jagged.

Case presentation:

Figure 35 A and B are taken with the immersion technique from the same eye. Note the presence of jagged and small corneal and retinal spikes, and the absence of the posterior lens spike in Figure 5A due to a large off-axis ultrasound direction resulting in an unrecordable measurement. In Figure 35 B the ultrasound beam is perpendicular to the major ocular surfaces, resulting in an excellent ultrasound display.

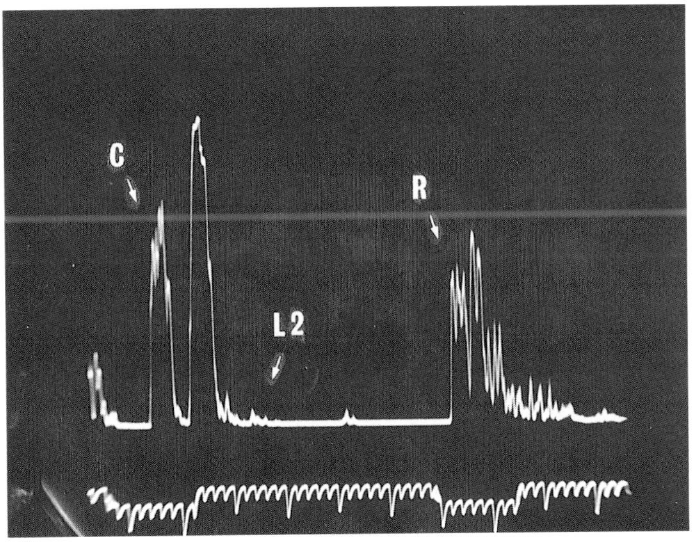

Figure 35 (A): A-scan display of an off-axis measurement taken with the immersion technique. Note the presence of poor corneal (C) and retinal (R) spikes and the absence of the posterior lens spike (L2).

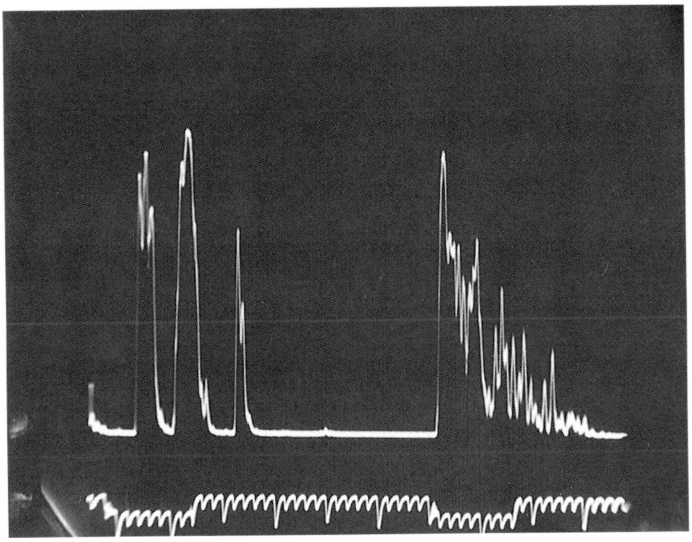

Figure 35 (B): A-scan display of the same eye with an on-axis measurement.

Wrong Placement of the electronic gates

The wrong placement of the gates can read the measurement between two unrelated surfaces instead of between the cornea and the retina.

Case presentation:

Figures 36 A and B are taken with a contact technique from the same eye. In Figure 36 A, the corneal gate was inadvertently moved to the right, resulting in a measurement of 19.98 mm in this emmetopic eye. This measurement was taken between the anterior surface of the lens and the anterior surface of the retina. In Figure 36 B, the corneal gate was moved back to its original position and the correct measurement of 22.88 mm was obtained.

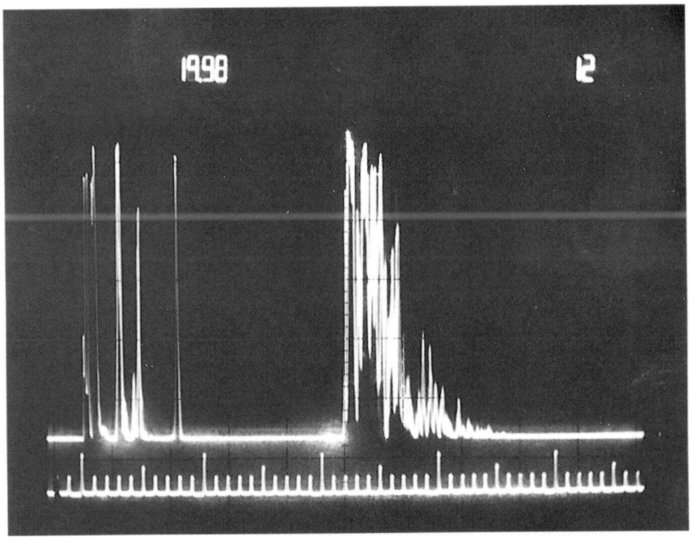

Figure 36 (A): A-scan pattern of a phakic eye where the overlight corneal gate is inadvertently moved to the right, yielding a measurement of 19.98mm from the anterior lens surface to the retina.

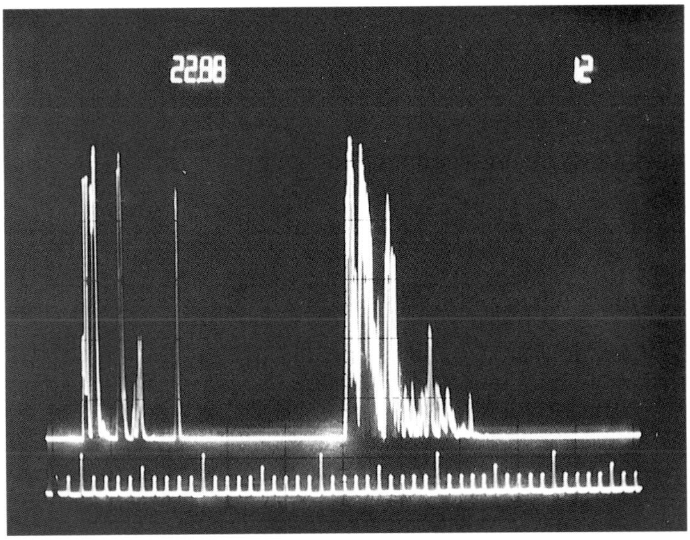

Figure 36 (B): A-scan pattern of the same eye where the overlight corneal gate has been repositioned over the corneal echospike with a correct measurement of 22.88mm.

Presence of Asteroid hyalosis

The presence of asteroid hyalosis(8) generates echospikes within the vitreous cavity that can be confused by the biometer as the retinal surface.

Case presentation:

Figure 37 A and B are taken with a contact technique from the same eye with asteroid hyalosis. In Figure 37 A, the large retinal gate picked up the presence of the calcified bodies within the vitreous, resulting in a measurement of 19.91 mm.

In Figure 37 B, the system sensitivity was decreased by 10 decibels and the retinal gate was repositioned to avoid the vitreous pathology, resulting in the correct measurement of 23.88 mm.

By decreasing the biometer's system sensitivity, the amplitude of all the echospikes will decrease to a point where the weaker vitreous spikes almost disappear. Also, when in doubt as to the nature of the vitreous pathology, a B-scan ultrasound can be helpful.

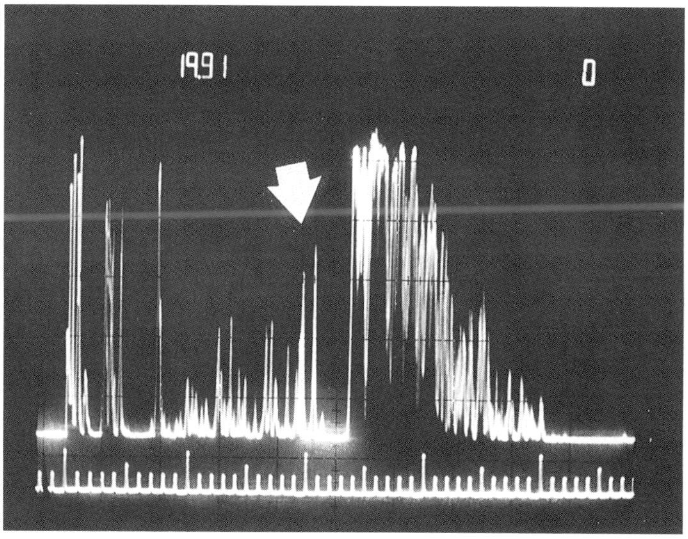

Figure 37 (A): A-scan pattern of a phakic eye with asteroid hyalosis. The large overlight retinal gate mistook the posterior asteroid bodies (arrow) for the retina, yielding measurement of 19.91mm.

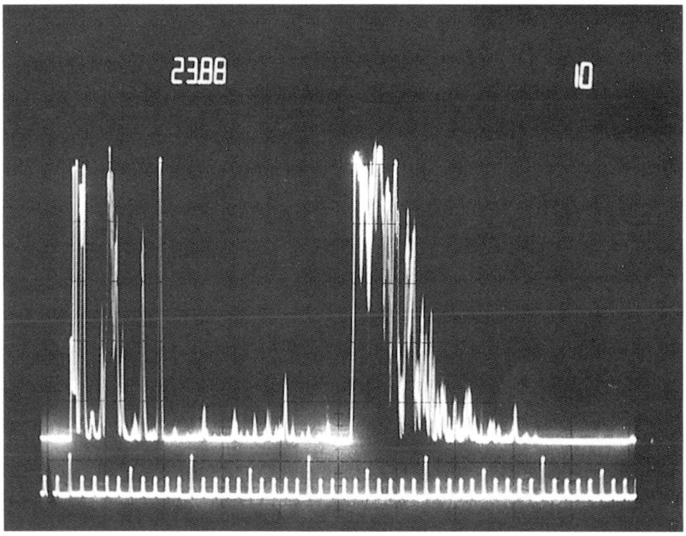

Figure 37 (B): A-scan pattern of the same eye with decreased system sensitivity and a repositioned retinal gate, yielding a correct measurement of 23.88mm.

2. LONGER AXIAL LENGTH MEASUREMENT

Longer measurements of the axial length occur in the presence of a pre-corneal echospike, a fluid bridge between the probe and the cornea, a poor retinal echospike or the use of an inaccurate velocity.

Entering a longer measurement of the AL in IOL power calculations will call for the use of a weaker IOL than is actually required, resulting in an induced hyperopia in the final post-operative refraction.

Pre-corneal echospike

A pre-corneal echospike is usually generated by an air bubble within the water-filled contact probe or within the scleral shell during an immersion technique(5).

Case presentation:

Figure 38 A represents an axial length measurement taken in the presence of a pre-corneal air bubble. The axial length was inaccurately read as 24.7 mm (Fig. 38A). Removal of the air bubble yielded the correct measurement of 22.8 mm (Fig. 38B).

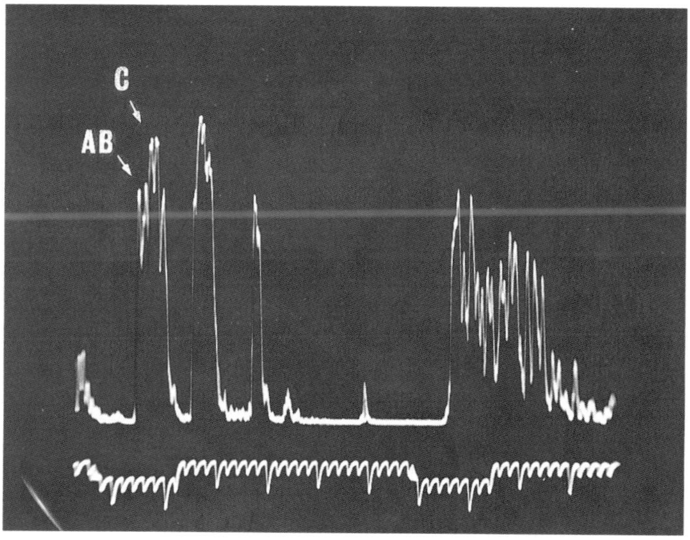

Figure 38 (A): A-scan pattern of a phakic eye measured in the presence of an air bubble (AB) in front of the cornea (C).

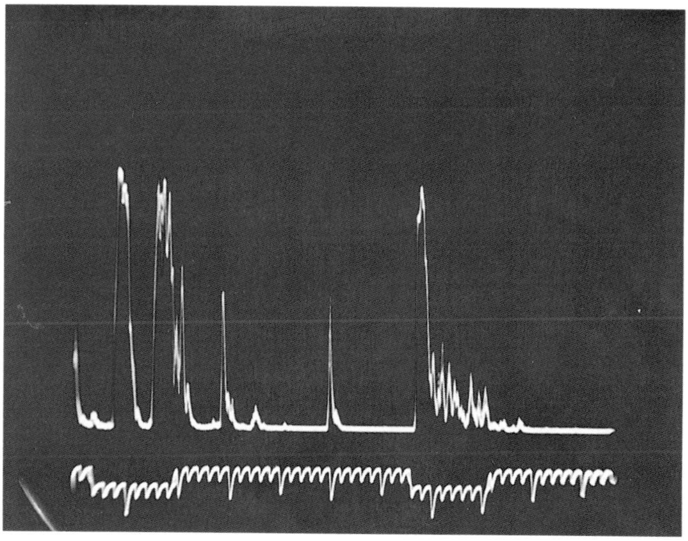

Figure 38 (B): A-scan pattern of the same eye after the air bubble has been removed.

Fluid bridge between the probe and the cornea

A tear film always lays between the probe and the cornea. An excessive tear layer or the presence of another conductive material such as methylcellulose will cause a fluid bridge and results in a longer measurement of the axial length(9).

Case presentation:

Figures 39 A and B are taken with the contact technique from the same eye. In Figure 39 A, the axial length was measured as 22.56 mm immediately after retinal fundus examination with a three-mirror gonioprism using 2.5% methylcellulose.

In Figure 39 B, the axial length was remeasured half an hour later and after washing the eye with balanced salt solution; the axial length was then correctly measured at 22.42 mm.

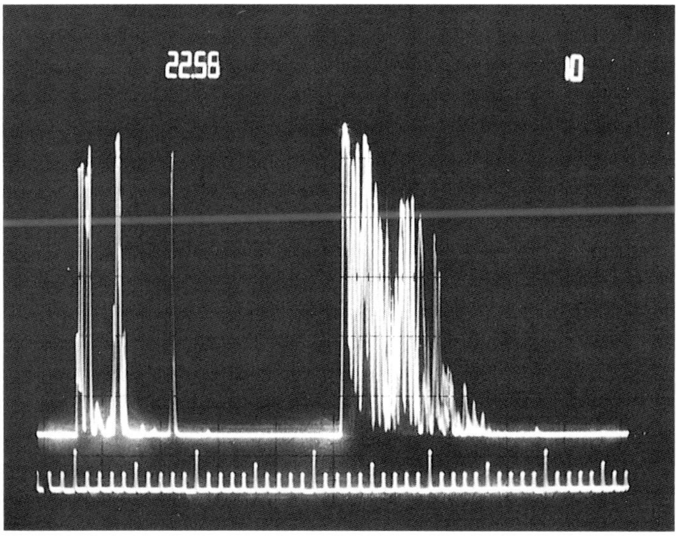

Figure 39 (A): A-scan pattern of a phakic eye with an excessive fluid bridge between the cornea and the probe, yielding a measurement of 22.56mm.

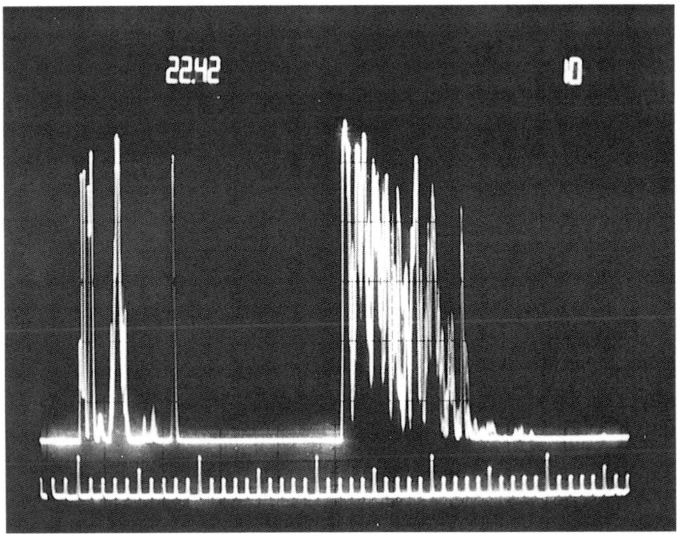

Figure 39 (B): A-scan pattern of the same eye after clearing the excessive fluid layer with a correct measurement of 22.42mm.

Poor retinal echospike

An off-axis measurement can yield a poor retinal echospike. The biometer will miss the retinal spike and read a longer measurement between the corneal and the scleral spikes(10).

Case presentation:

Figures 40 A and B are taken from the same eye. In Figure 40 A the retinal spike is weak yielding a false measurement of 23.59 mm.

In Figure 40 B, the retinal spike is sharp and steeply rising with a correct measurement of 22.44 mm.

Inaccurate velocity

An inaccurate velocity can be inadvertently used when measuring an aphakic eye(11).

Case presentation:

An aphakic eye is measured prior to secondary lens implantation. The technician forgot to change the preset sound velocity of 1552 m/s yielding a measurement of 23.41 mm. When the accurate sound velocity of 1534 m/s was used, the correct measurement of 23.13 mm was obtained.

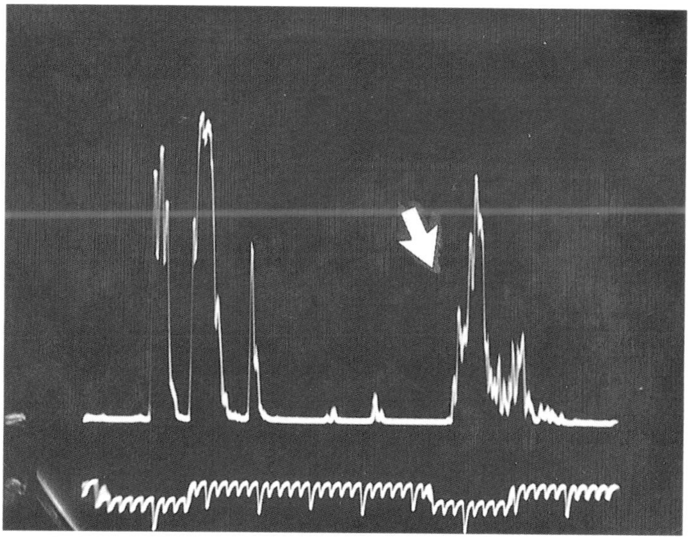

Figure 40 (A): A-scan pattern of a phakic eye with a weak retinal spike (arrow)

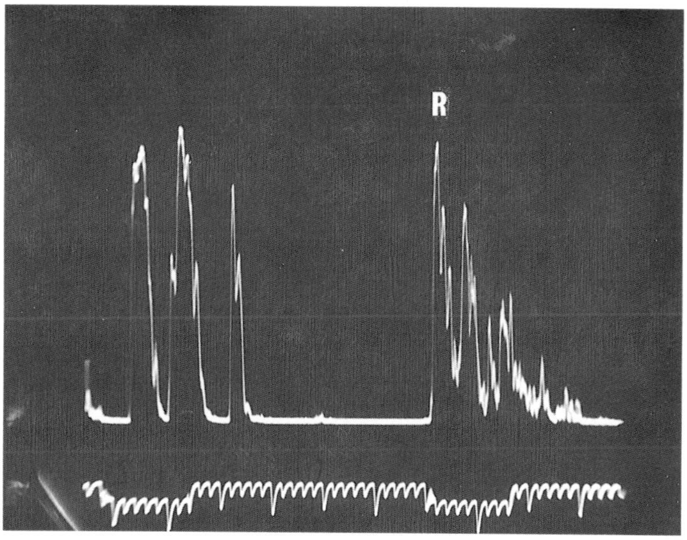

Figure 40 (B): A-scan pattern of the same eye with a correct display of a sharp and steeply rising retinal spike (R).

3. *DETECTING SIGNIFICANT INTRAOCULAR PATHOLOGY*

Besides the vitreous pathology previously described, there are cases where the ultrasound pattern is difficult to interpret. In such cases, a B-scan and/or standardized A-scan examination will make the correct diagnosis (12-14). The surgeon will then decide whether the cataract surgery is indicated or not, and whether an intraocular implant should be used if the cataract is removed.

In most cases, the pathology is uniocular, and cannot be visualized due to the mature or almost mature cataract.

The prephthisical eye

In a prephthisical eye, the globe is relatively soft and has decreased in size following trauma. Sometimes these changes are very subtle and can be missed on examination.

Case presentation:

This 32-year-old man has a mature cataract in his right eye with vision of light perception. There is a five-year history of a trauma to that eye. Measurements from both eyes reveal an axial length of 19.31 mm in the right eye (Fig. 41A) and of 23.87 mm in the left eye.

A B-scan (Fig. 41B) shows the presence of diffuse choroidal thickening in the traumatized eye, characteristic of a prephthisical condition.

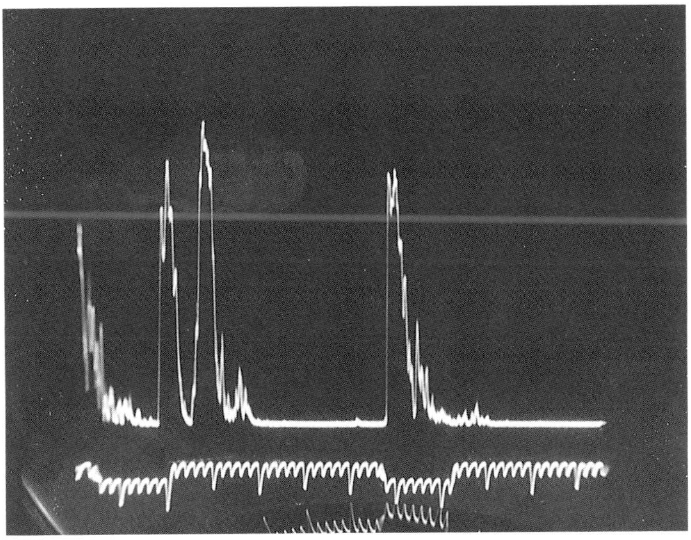

Figure 41 (A): Axial length measurement of a prephthisical eye (19.31mm).

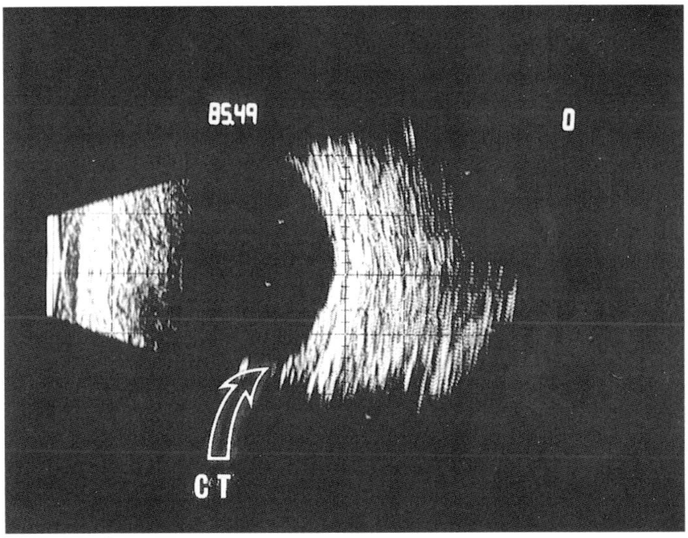

Figure 41 (B): B-scan pattern of the prephthisical eye with choroidal thickening (CT).

Retinal detachment

In an uncomplicated retinal detachment, the retina is totally separated from the pigment epithelium, but remains attached to the optic disc and to the ora serrata.

Case presentation:

This 72-year-old lady has had a decreased vision in both eyes, more so in the right eye for the past three years. Visual acuity is light perception with a mature cataract.
Measurement of the axial length was very difficult to perform (Fig. 42A) and showed the presence of an extra spike in front of the posterior segment layers. The B-scan ultrasound revealed a shallow retinal detachment (Fig. 42B).

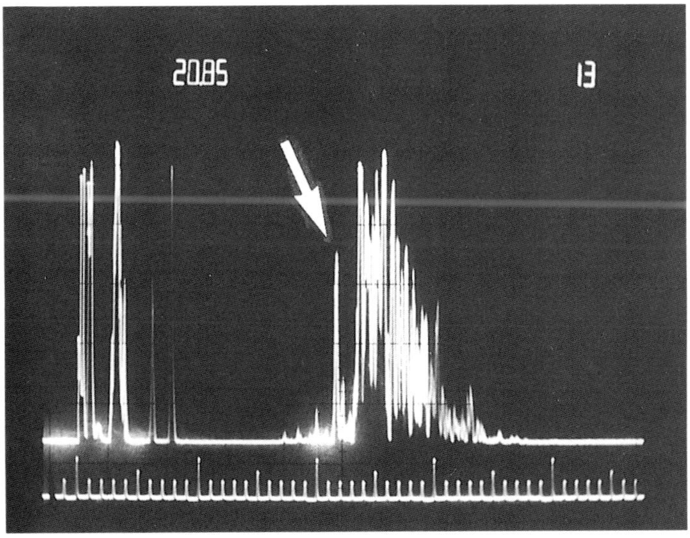

Figure 42 (A): A-scan pattern of a phakic eye with retinal detachment, where the retinal spike (arrow) is separate from the posterior layers of the eye.

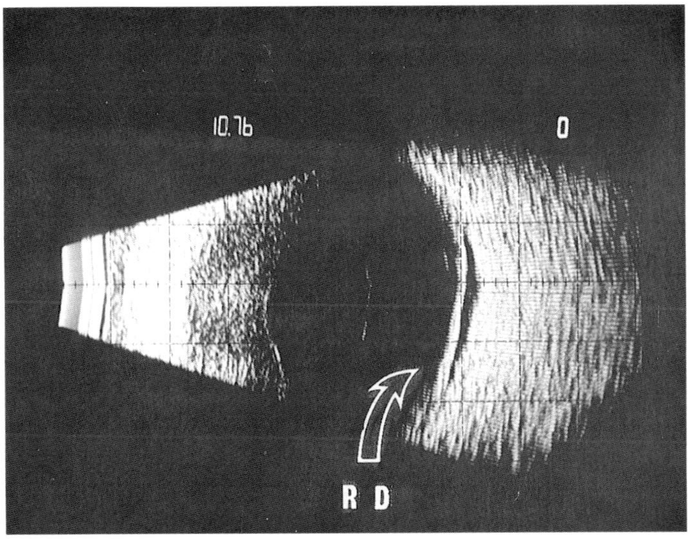

Figure 42 (B): B-scan pattern of the same eye, showing the detached retina (RD).

Disciform macular degeneration

A disciform macular degeneration is caused by subretinal scar tissue formation secondary to bleeding. The choroid is elevated and, in certain cases, it can simulate a choroidal tumor.

Case presentation:

This 65-year-old lady had an immature cataract in the left eye with nuclear sclerosis and very dense posterior subcapsular precipitates, making fundus examination impossible. The fellow eye had a 20/40 vision, minimal cataract changes and few macular drusens.

Measurement of the axial length was difficult and yielded a measurement of 20.85 mm while the other eye measured 23.20 mm. B-scan revealed the presence of a macular degeneration, subretinal hemorrhage and a diffuse vitreous hemorrhage (Figure 43).

Choroidal mass

A choroidal mass elevates the overlying retina, causing an axial length shortening. The differential diagnosis of a choroidal mass includes choroidal melanoma, hemangioma or a metastatic carcinoma.

Case presentation:

This 76-year-old lady had a dense cataract in the right eye, increased intraocular pressure and a vision of hand motion. The fellow eye had a 20/80 cataract and a normal pressure.

Measurement of the axial length was difficult and suspicious of intraocular pathology. B-scan revealed the presence of a choroidal mass, highly suggestive of a choroidal melanoma. (Fig. 44).

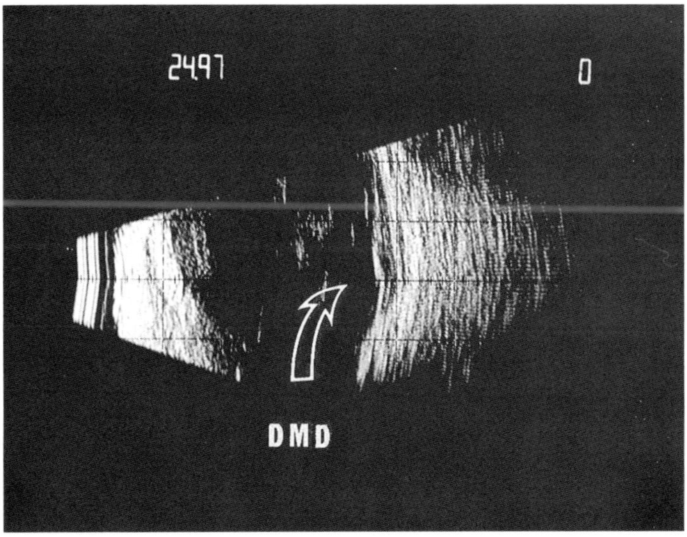

Figure 43: B-scan pattern of a phakic eye with disciform macular degeneration (DMD), showing the flat echogenic subretinal elevation caused by bleeding and scarring.

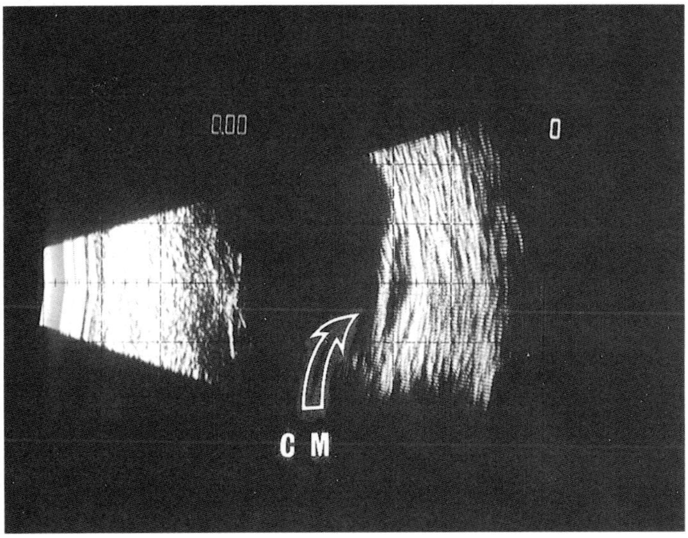

Figure 44: B-scan pattern of a phakic eye showing the presence of a dome-shaped choroidal elevation, highly suggestive of a choroidal melanoma (CM).

Vitreous hemorrhage

In the presence of a vitreous hemorrhage, the red blood cells are dispersed within the vitreous cavity. The hemorrhage usually is limited posteriorly by the posterior vitreous layer.

Case presentation:

This 57-year-old diabetic patient had a cataract in the right eye and a vision of counting fingers. The left eye had 20/30 vision, minimal cataract changes and proliferative diabetic retinopathy. Measurement of the axial length revealed a relatively normal A-scan pattern (Fig. 45A) with multiple small echospikes in the vitreous cavity. B-scan was performed on the right eye and revealed diffuse vitreous hemorrhage (Fig. 45B). The red blood cells in the vitreous are very low reflective and, since the system sensitivity is normally decreased for AL measurement, the hemorrhage can be easily missed.

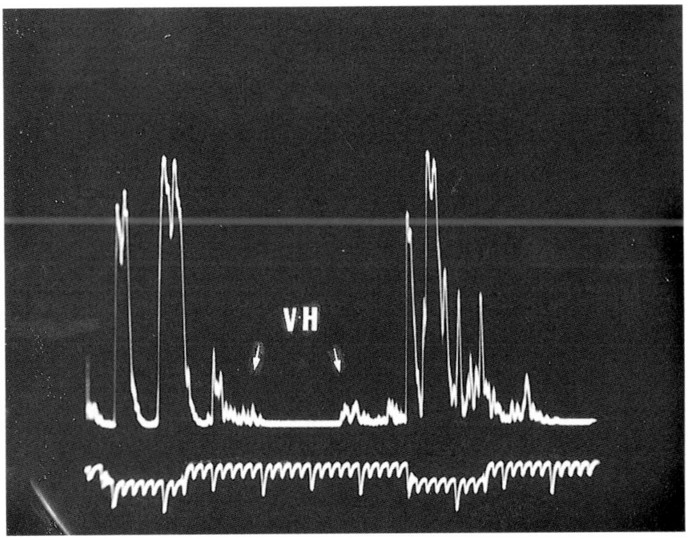

Figure 45 (A): A-scan pattern of a phakic eye with vitreous hemorrhage. Note the presence of multiple, small echospikes (VH) caused by the dispersed red blood cells in the vitreous cavity.

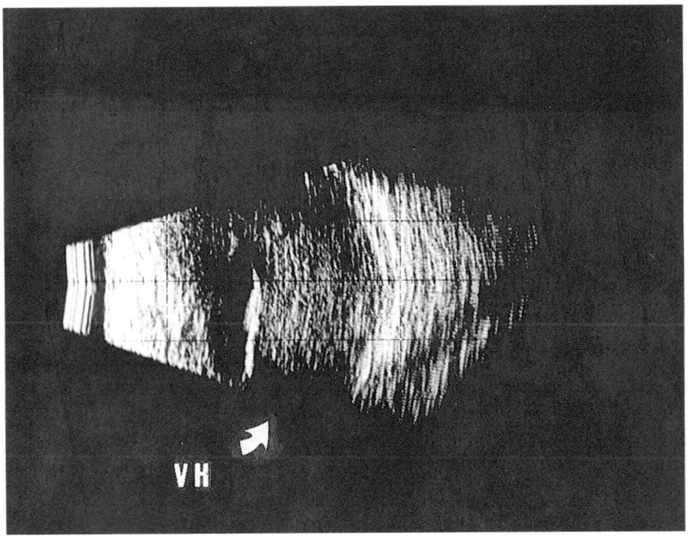

Figure 45 (B): B-scan pattern of the same eye showing a diffuse vitreous hemorrhage (VH).

Posterior pole staphyloma

A posterior pole staphyloma usually is present in severe axial myopia. Occasionally it can be undetected if associated with a mature cataract in an eye with unilateral axial myopia(12).

Case presentation:
This 48-year-old male is referred by his ophthalmologist for repeat AL measurement and IOL calculations because of difficulties in measurement. An immersion technique was used and yielded a measurement of 28.80 mm in the cataractous eye (Fig. 46A). B-scan examination shows the increased curvature of the posterior pole in the cataractous eye (Fig. 46B) with a central staphyloma.

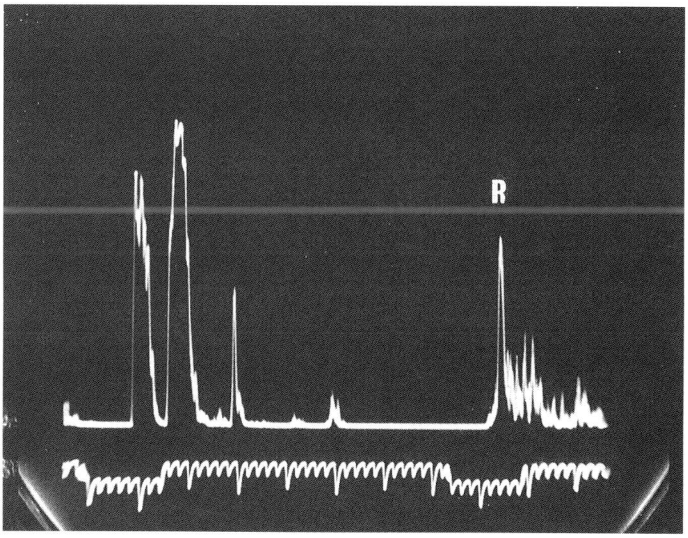

Figure 46 (A): A-scan pattern of a phakic eye with a posterior pole staphyloma, yielding a measurement of 28.80mm. Note the presence of a weak retinal spike (R) due to the increased curvature of the posterior pole.

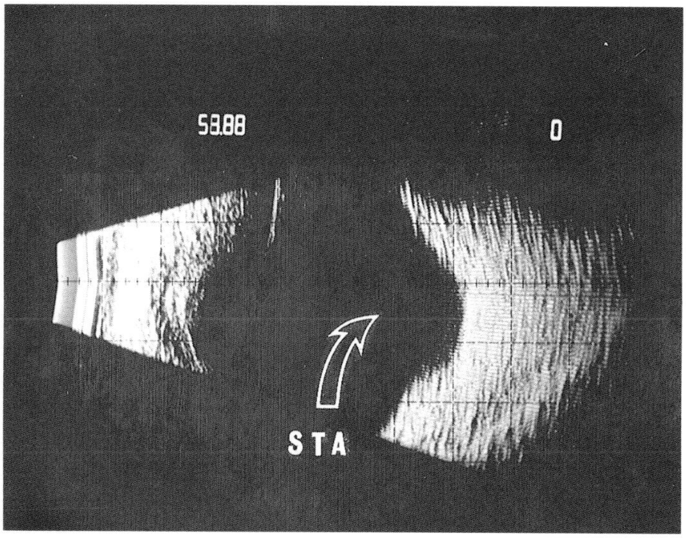

Figure 46 (B): B-scan pattern showing the increased curvature of the posterior pole caused by the staphyloma (STA).

4. RECOMMENDATIONS

There is no fool-proof method to avoid an error in axial length measurement, but I would like to share a protocol that has proven to be very effective(15).

1. Bilateral axial length measurements are performed by a certified technician. Each eye is measured three times and pictures or print-outs are obtained. Consistency is very important. Variations in the ultrasound probe design, in the sound velocity, or in the technique can cause up to 0.5 mm difference in measurement.

2. Prior to surgery, the pictures of the measurements are reviewed. If they are not satisfactory, the patient is recalled and the eye is remeasured. When needed, a B-scan examination will rule out intraocular pathology.

3. The ultrasound results are then correlated with the clinical data and a choice of an implant is made. In most cases, both eyes are within 0.5 mm of each other and the ultrasound results correlate with the clinical data; hypermetropes usually measure less than 23 mm, emmetropes between 23 mm and 24 mm, and true myopes over 24 mm. However, discrepancies will be encountered in some cases. The surgeon should then use his or her clinical judgment in analyzing the data, because it is ultimately his or her responsibility. If the surgeon is not satisfied with the calculations or suspects a possible error, he or she should have the measurements repeated, either under his or her supervision, or by an outside consultant. It is then, and only then, that a potential "9 diopter" surprise can be avoided.

REFERENCES - CHAPTER VI

1. Ossoinig, KC, Byrne, SF and Weyer, NJ: Standardized echography, Part II: Performance of standardized echography by the technician, Int.Ophthalmol. Clin. 19:283-285, 1979.

2. Shammas, HJF: Manual versus electronic measurement of the axial length. In Ultrasonography in ophthalmology, pp. 225-229, Proceedings of the 1982 Ninth SIDUO Congress, J.S. Hillman and M.M. LeMay (eds.), The Hague, 1983, Dr. W. Junk Publishers.

3. Salz, JJ, Reader, AL III: Lens implant exchanges for incorrect power: Results of an informal survey. J. Cataract Refract. Surg. 14:221-224, 1988.

4. Binkhorst, RD: The accuracy of ultrasonic measurement of the axial length of the eye. Ophthalmic Surgery. 12:363-365, 1981.

5. Shammas, HJ: Atlas of Ophthalmic Ultrasonography and Biometry. pp. 273-308. The C.V. Mosby Co., St. Louis, 1984.

6. Bonnac, JP, Massin, M, and Poujol, J: Comparison between the measures of length of a myopic eye in the axis and in the para-axial regions. In Ultrasonography in ophthalmology, J. Francois and F. Goes (eds.), pp. 283-286, Basel, Switzerland, 1975, S. Karger.

7. Shammas, HJ: Ultrasound diagnosis in unilateral axial myopia. In "Ophthalmic Ultrasonogaphy". Proceedings of the 12th SIDUO Congress. R. Sampaolesi, Editor. Kluwer Academic Publishers. Documenta Ophthalmologica. 53:149-153, 1990.

8. Martin, RG, Safir, A: Asteroid hyalosis affecting the choice of intraocular lens implant. J. Cataract Refract. Surg. 13:62-64, 1987.

9. Kendall, CJ: Ophthalmic Echography. pp. 57-106, The Ophthalmic Technical Series. 1990, Slack, Inc., Thorofare, NJ.

10. Ossoinig, KC: Standardized echography: basic principles, clinical applications, and results, Int. Ophthalmol. Clin. 19:127-210, 1979.

11. Giers, V, Epple, C: Comparison of A-Scan device accuracy. J. Cataract Refract Surg. 16:235-242, 1990.

12. Shammas, HJ and Milkie, CF: Mature cataracts in eyes with unilateral axial myopia. J. Cat. Refr. Surg. 15:308-311, 1989.

13. Tjan, TT: Ultrasonic diagnosis of ocular pathology: its significance for cataract surgery with and without IOL implantation. J.Am. Intraocular Implant Soc. 5:15-21, 1979.

14. Wainstock, MA: Ultrasonography: its role in the success of intraocular implant surgery. Int. Ophthalmol. Clin. 19:43-50, 1979.

15. Shammas, HJ: The 9 diopter surprise revisited. J. Cat. Refract. Surgery 14: 580, 1988.

CHAPTER VII

MEASURING THE CORNEAL POWER

The corneal power is the second most important factor affecting IOL power calculations after the axial length. A keratometric error of 1 diopter affects the postoperative refraction by approximately the same amount.

1. MANUAL KERATOMETRY

Manual keratometry is the most commonly used method to measure the corneal curvature(1). It is fast, easy, cheap and very accurate in most average cases.

The patient is seated behind the keratometer with the chin well positioned in the chin rest and the forehead in touch with the head rest (Figure 47). The keratometer is directed towards the eye to be examined while the occluding shield is placed in front of the opposite eye. The keratometer is focused on the central portion of the cornea using the focusing knobs. At this time, the central ring appears as one circle. The instrument is rotated to align the (-) signs in the same vertical meridian and the (+) signs in the same horizontal meridian. This will determine the axis of any pre-existing astigmatism. The left drum is rotated to superimpose the (+) signs and a horizontal measurement is taken (Fig. 48). The right drum is then rotated to superimpose the (-) signs and a vertical measurement is taken (Fig. 49).

IOL power formulas rarely use the horizontal (K1) and the vertical (K2) readings. They usually call for the average value (K):

$$K = 0.5 (K1 + K2)$$

It is important to remember that the keratometer has to be calibrated every 6 months. A set of calibration balls with a known radius of curvature can be used. The horizontal and vertical drums are then set at the same value.

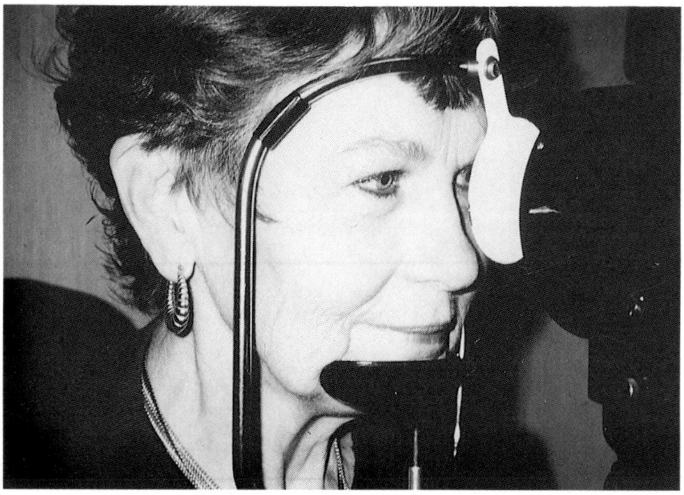

Figure 47: The patient is seated behind the manual keratometer and the occluding shield is placed in front of the opposite eye.

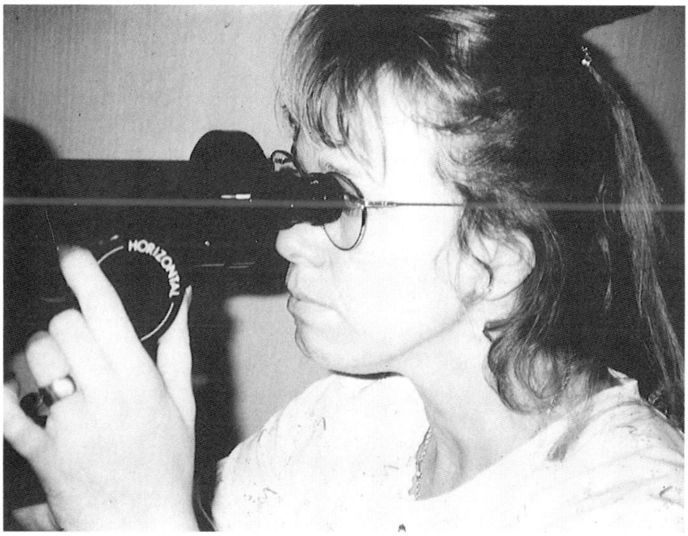

Figure 48: The technician rotates the left drum to obtain a horizontal measurement.

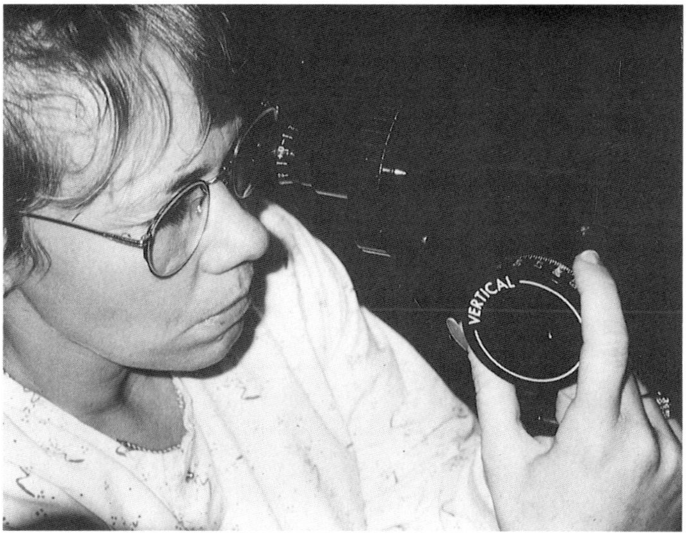

Figure 49: The technician rotates the right drum to obtain a vertical measurement.

2. AUTOMATIC KERATOMETRY

An automatic keratometer is designed to give accurate, objective and reproducible measurements of the K readings(1).

The technique consists of seating the patient behind the automatic keratometer with the chin resting in the chin cup and forehead against the forehead rest.

The eye to be examined is selected by pressing the "R eye" and "L eye" buttons. The patient is instructed to look at the center of the red fixation light and to hold the head steady. Using the control ball, the flashing green light is focused on the central cornea between the two smaller guidelights until it stops flashing and glows continuously. The "READ" button is pressed and a measurement is taken.

A print-out is obtained for safe keeping in the patient's record. It provides the K readings from the central cornea in the horizontal and in the vertical meridians; the readings are obtained in diopters and in millimeters. Additional information not required for IOL power calculations include the amount of total astigmatism, the apical K reading, a shape factor to determine if the cornea is spherical or more elliptical, and a mire index that compares the measurements to a model eye.

3. CORNEAL TOPOGRAPHY

A corneal analysis system is used to evaluate central and peripheral corneal curvature(2,3). The patient is seated behind a corenoscope that projects a 16-ring conical placido disc on the cornea. A small camera captures the image reflection and transmits it to a computer that converts the data into a series of color graphic displays.

Although a regular keratometer gives accurate and reproducible measurements in most cases, corneal topography can be helpful in certain cases:

- -When the cornea is flatter than 40 diopters or steeper than 46 diopters.
- -When the surgeon wants to better evaluate the pre-existing astigmatism.
- -When the cornea is irregular, i.e., after trauma or in the presence of a concomitant keratoconus.
- -When some type of corneal surgery has already been performed, i.e., radial keratotomy or laser photorefractive keratectomy.

4. *THE CORNEAL POWER FOR A CONCURRENT ASTIGMATIC KERATOTOMY*

Concurrent astigmatic keratotomy and implant surgery are being performed successfully to reduce the pre-existing astigmatism. Corneal topography will detect the correct amount of astigmatism and its axis prior to surgery(3-5)

Astigmatic keratotomy flattens the steep corneal meridian. However, this flattening is associated with a slight to moderate steepening of the cornea 90 degrees away. This is known as the coupling effect. If the amount of flattening is equal to the amount of steepening 90 degrees away, then the average K readings remain the same, and the power of the IOL remains the same. In fact, the amount of flattening slightly exceeds the amount of steepening, ending in a flatter cornea, which affects the IOL power calculations.

Ideally, the cornea is remeasured after the keratotomy and the IOL power is recalculated. However, if the microscope is not equipped with a keratometer or if the surgeon does not want to redo the calculations in the operating room, then the expected corneal curvature has to be calculated. I suggest subtracting 0.25 diopters from the average preoperative K readings for every diopter of astigmatism to be corrected.

Clinical application:

Preoperative K readings: 41,000/45.00 diopters
Average preoperative K: 43.00 diopters
Plan: Astigmatic keratotomy to correct 4.00 diopters
Expected postoperative K: 43.00-(0.25 x 4) = 42.00 diopters
This last value of 42.00 diopters is used in the IOL power calculations.

5. *CORNEAL POWER AFTER RADIAL KERATOTOMY AND LASER PHOTOREFRACTIVE KERATECTOMY*

Calculating the IOL power after radial keratotomy(RK) is more difficult than in normal eyes due to the changes that have occurred to the cornea and the K readings do not accurately reflect the true power of the cornea(6,7). Often, after the RK, there is a discrepancy between the myopic correction achieved and the change in the keratometric readings.

Clinical Application:

An eye with K readings of 45.00 diopters before RK had 5 diopters of myopia corrected at the cornea. After the RK, one would expect readings of 40 diopters; instead the keratometer gives an average reading of 41.5 diopters. The question becomes which "K readings" have to be entered into the IOL power formula prior to the cataract surgery?

Different methods are available to measure the corneal curvature after radial keratotomy. An easy way of doing it is by using the "refractive history" method: Start with the K readings of the eye prior to RK and subtract from it the amount of myopia corrected by the procedure at the cornea (RC) whereby

$$RC=Rs ./. (1-0.012\,Rs).$$

In this formula, Rs represents the amount of myopia present in the spectacle prescription; if there is any astigmatism, the myopic correction is replaced by the spherical equivalent value. This method works best if the same surgeon has performed the RK and has followed the patient post-operatively. It is important not to include in the calculations any myopic changes caused by the cataract itself. Another drawback with this method is that the myopic correction achieved by radial keratotomy can change with time and it is not unusual to notice a hyperopic shift a few years after the RK.

A more elaborate and more difficult way of measuring the corneal curvature is the "contact lens method": apply a hard contact lens that has the patient's refraction and a known base curve over the cornea and perform an over refraction; any additional myopia is subtracted from the contact lens base curve to obtain the true corneal base curve. Errors occur due to refractive changes caused by the tear film layer (between the contact lens and the cornea) and by the cataract itself.

Keratometers use concentric rings that are projected on the cornea. Ideally, these rings have to be extremely thin and should not exceed 0.3 mm in diameter. The accuracy of the keratometric readings on the corneal vertex is directly related to the diameter of the corneal image projected within these rings. Furthermore, after radial keratotomy, any reading at the corneal vertex might not reflect the actual corneal curvature; instead, average keratometric readings are needed from the entire functional optical zone of the cornea which average at least 3.5 mm. To that end, corneal topography units are much more accurate than the usual keratometers and that would be the method of choice in measuring the "K readings" prior to the cataract surgery in patients who have had radial keratotomy.

The same principles apply to patients who have had laser photorefractive keratectomy; corneal topography will be more accurate than any manual keratometer in establishing the correct K readings(8,9).

6. CORNEAL POWER IN THE COMBINED PROCEDURE OF CORNEAL TRANSPLANT, ECCE AND LENS IMPLANT

In the presence of a cataract with marked endothelial dystrophy, causing corneal edema(1), the surgeon has three choices:

- Removing the cataract first, followed by a corneal transplant. This procedure can be done if the cornea is still relatively clear. Variations in the final refraction can be encountered, depending on the corneal transplant dioptric power.
- Performing a corneal transplant first, followed a few months to a year later by cataract and implant surgery. The final refraction is more predictable because the power of the corneal transplant is known at the time of cataract surgery and any induced astigmatism from the corneal transplant can be corrected during the second surgery. However, this scheme requires two surgeries and a long period of visual rehabilitation.
- With combined corneal transplant, ECCE, and lens implant surgery, the final refraction is not as predictable, but it eliminates a second surgery, offers a faster visual recovery, and avoids additional risks to the graft's endothelium.

With this triple procedure there is a lack of accuracy in the IOL power calculations due to the unknown power of the new cornea(10,11). The power of a cloudy cornea can be difficult to measure preoperatively. Furthermore, such a measurement can be inaccurate; a scarred cornea will yield weaker K readings. In such cases, both eyes should be measured, and if there is a large discrepancy between the two, it is best to use the K readings of the opposite cornea. Also the surgeon should remember that a corneal button 0.2 mm larger than the recipient's bed avoids any major myopic inducement. If the button is 0.2 to 0.5 mm larger than the recipient's bed, it steepens the central cornea, causing a myopic shift of 1.00 to 2.00 diopters.

REFERENCES - CHAPTER VII

1. Shammas, HJ: Atlas of Ophthalmic Ultrasonography and Biometry. pp. 273-308. The C.V. Mosby Co., St. Louis, 1984.

2. Rowsey, JJ, Reynolds, AE, Brown, R: Corneal topography: corneascope. Arch. Ophthalmology. 99:93-100, 1981.

3. Koch, DD and Haft, EA: Introduction to corneal topography. In "An Atlas of Corneal Topography". pp. 1-30; Sanders, DR and Koch, DD, editors, Slack, Inc. Publishers, Thorofare, N.J.

4. Axt, JC: Longitudinal study of postoperative astigmatism. J. Cataract Refract Surg. 13:381-388, 1987.

5. Stainer, GE, Binder, PS, Parker, WT, and Perl T: The natural and modified course of post-cataract astigmatism. Ophthal. Surg. 13:822-827, 1982.

6. Holladay, JT: IOL calculations following radial keratotomy surgery. Refractive and Corneal Surgery 5:36, 1989.

7. Koch, DD, Liu, JF, Hyde, LL, Rock, RL and Emery, JM: Refractive complications of cataract surgery after radial keratotomy. Amer. J. Ophthalmol. 108:676-682, 1989.

8. Colliac, JP, Shammas, HJ and Bart, DJ: Photorefractive keratectomy for the correction of myopia and astigmatism, Am. J. Ophthalmol. 117:369-380, 1994.

9. Colliac, JP, Shammas, HJ and Bart, DJ: Reply to Schipper I: Correspondence. Am. J. Ophthalmol 118:41, 1995.

10. Lindstrom, RL, Harris, WS, Doughman, DJ: Combined penetrating keratoplasty, extracapsular extraction and posterior chamber lens implantation. AIOIS J. 7:130-132, 1981.

11. Binder, PS: Secondary intraocular lens implantation during or after corneal transplantation, Am J. Ophthalmology. 99:515-520, 1985.

CHAPTER VIII

THE ANTERIOR CHAMBER CONSTANT

All original theoretical formulas take into consideration the anterior chamber depth (ACD) defined as the distance between the corneal vertex and the anterior surface of the IOL. In these formulas, the ACD has been a measured distance and a fixed value within the formula (1-3).

The advent of the SRK regression formula brought with it the concept of the "A" Constant. The A Constant varied not only with the anterior chamber depth, but also with a variety of factors that had not been taken into consideration in the original theoretical formulas, namely:

- Variations in the axial length measurement due to different biometers, different velocities and different measurement techniques.
- Variations in the IOL design
- Variations in the surgical techniques.

The concept of individualizing the A Constant revolutionized IOL power calculations leading to the new generations of IOL formulas.

Three constants are used in modern IOL formulas:

- The ACD Constant representing the postoperative anterior chamber depth estimate in Binkhorst, Hoffer and Shammas formulas
- The A Constant in the SRK II and SRK-T formulas
- The S Constant in the Holladay formula

1. The ACD Constant

By definition, the anterior chamber depth is the distance taken postoperatively between the corneal vertex and the anterior surface of the IOL(4,5). With ultrasound (Fig. 50), the anterior chamber depth can be precisely measured between the anterior surface of the cornea (C) and the anterior surface of the pseudophakos lens (P). Newer units will give the reading automatically.

It is important to note that the postoperative anterior chamber depth does not correlate with the preoperative anterior chamber depth. Instead, it correlates with the placement of the IOL whether it's in the anterior chamber, in the sulcus or in the capsular bag.

The ACD Constant also varies with the implant's configuration and the location of its optical center. The use of a meniscus lens with its anteriorly located optical center calls for a smaller ACD Constant than a biconvex IOL where the optical center is more posteriorly located.

The ACD Constant has a different value in each of the formulas where it is used because of the different characteristics of these formulas. When referring to a specific ACD, it becomes important to refer to it as the Binkhorst ACD, the Hoffer ACD or the Shammas ACD.

The commonly used ACD values for the different IOL styles are as follows:

Anterior chamber lens	2.8 to 3.1 mm
Iris-supported lens	3.3 to 3.5 mm
Posterior chamber lens in the sulcus	
Plano-convex	3.7 to 4.0 mm
Biconvex	3.8 to 4.1 mm
Posterior chamber lens in the bag	
Plano-convex	4.3 to 4.5 mm
Biconvex	4.8 to 5.1 mm

An error of 1 mm in the ACD Constant affects the postoperative refraction by approximately 1.0 D in a myopic eye, 1.5 D in an emmetropic eye and up to 2.5 D in a hyperopic eye.

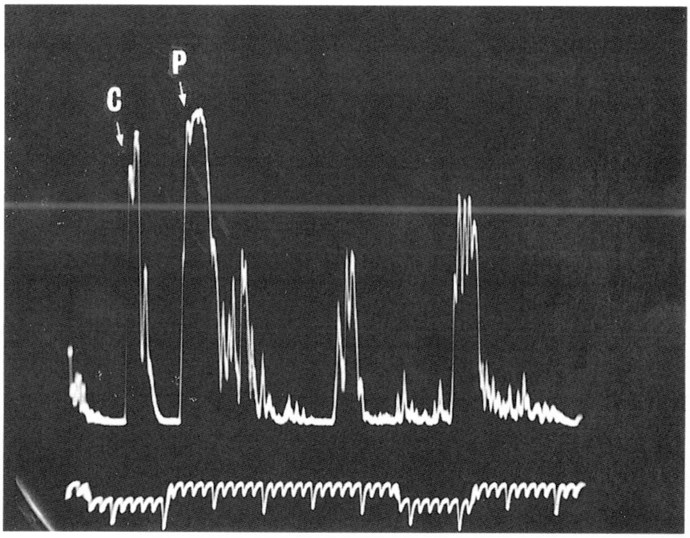

Figure 50: A-Scan pattern of a pseudophakic eye. The postoperative anterior chamber depth is measured between the anterior corneal surface (C) and the anterior surface of the pseudophakos lens (P).

2. The A Constant

The A Constant has been designed originally for the SRK linear equation: $P = A - 2.5L - 0.9K$

Where P is the power for emmetropia in diopters, L is the axial length in mm and K is the corneal power in diopters. The A Constant encompasses multiple variables that include the implant manufacturer, implant style, surgeon's technique, implant placement within the eye and measuring equipment.

Because of its simplicity, the SRK formula became very popular and the A Constant became the value used to characterize IOLs used in surgery. Unfortunately, the IOL power formulas are not that simple and that has lead to the development of the SRK II regression formula and then to the SRK-T theoretical formula. However, the same A Constant can be used with all three formulas(6,7).

The most commonly used A Constants are as follows:

Anterior chamber lens	115.0 to 115.3
Iris-supported lens	115.5 to 115.7
Posterior chamber lens in the sulcus	
Plano-convex	115.9 to 116.2
Biconvex	116.6 to 117.2
Posterior chamber lens in the bag	
Plano-convex	117.5 to 117.8
Biconvex	117.8 to 118.8

Clinical application:

Surgery is planned and IOL calculations call for a 21 diopter posterior chamber implant for emmetropia. During surgery the capsule breaks and vitreous is lost. The surgeon now has to insert an anterior chamber lens and no calculations for this implant have been made. A decision on the implant power has to be made on the spot. This is where the A Constant comes in handy no matter what formula has been used.

"A Constant" of the posterior chamber lens = 117.8

"A Constant" of the anterior chamber lens = 115.3

In most cases, the power of the IOL lens for emmetropia varies in a one to one relationship with the "A Constant"; if A decreases by 1, P decreases by 1 diopter also. This straight relationship adds to the simplicity and popularity of the A Constant.

In our case, the "A Constant" is lower by 2.5 diopters if an AC lens is used; then the implant power has to be decreased by the same amount to avoid any unwanted postoperative myopia, and an 18.5 diopter AC lens should be inserted to keep the eye in the emmetropic range.

3. The S Constant

The postoperative ACD depth which represents the distance between the corneal vertex and the optical plane of the IOL is actually the sum of two components:

- The anatomic ACD which is the distance from the corneal vertex to the anterior iris plane after surgery. This distance is more accurately predicted by using a mathematical formula based on the corneal curvature and the axial length.
- The distance between the anterior iris plane and the effective optical plane of the implanted IOL. This distance has been labeled as the surgeon's factor (S factor).

Although theoretically the S factor is a measurable distance, it should be personalized by solving the formula in reverse. Just like the A Constant, the S factor represents variations due to lens style, lens manufacturer, surgeon's technique and measurement devices(8,9).

Due to the close relationship between the S factor and ACD Constant (almost one to one), a change of 1 unit in the S factor is identical to a change of 1 mm in the ACD and affects the postoperative refraction in an average eye by around 1.5 diopters.

The most commonly used S factors are as follows:

Anterior chamber lens	-0.75 to -0.40
Iris supported lens	-0.10 to +0.10
Posterior chamber lens in the sulcus	
Plano-convex	0.1 to 0.3
Biconvex	0.4 to 0.7
Posterior chamber lens in the bag	
Plano-convex	0.9 to 1.1
Biconvex	1.2 to 1.6

4. INDIVIDUALIZING THE CONSTANTS

All three constants (ACD, A or S) have to be personalized to accommodate any consistent shift that might affect the IOL power calculations, namely the surgical technique, the biometer used, the implant's style and manufacturer.

The constant has to be back-calculated for a series of 20 or more cases. It is important for the series under review to have the same parameters:

- Same surgeon
- Same surgical technique:
  - . Limbal, scleral or corneal incision
  - . Planned extracapsular or phacoemulsification
  - . Capsulorhexis or can-opener anterior capsulotomy
  - . Implant fixation in the sulcus or in the bag
  - . Wound closure with radial, horizontal or with no sutures
  - . Postoperative steroid regimen
- Same biometry unit and same technician taking the measurement
- Same keratometer and same technician taking the readings
- Same IOL style from the same manufacturer

When the surgeon first uses a new IOL, it is best to use the designated constant, either from the manufacturer or from available tables. After having used 20 to 24 implants, the surgeon should personalize the constant by filling a data sheet similar to the one on the next page and back-calculating the constant for each case. It is best for the initial implants to be used in average eyes between 23 and 24 mm to obtain only minimal variation in the personalized constant. It is also best to use only cases with a stable refraction and a good visual acuity to avoid errors caused by concomitant intraocular diseases.

The constant (ACD, A or S) is then back-calculated for each case using available computer programs and calculators. The constant values are averaged to obtain the "Personalized Constant".

Data Sheet

Surgeon's name: ______________________ IOL Style: ______________

Surgeon's Technique: ______________________ IOL Manufacturer: ______________

#	Patient's Initials	Axial Length	Preop K	IOL Calculated	Expected Post-Op Refraction	Actual Post-OP Refraction	Back Calculated Constant
1							
2							
3							
4							
5							
6							
7							
8							
9							
10							
11							
12							
13							
14							
15							
16							
17							
18							
19							
20							
21							
22							
23							
24							
Average Constant =							

REFERENCES - CHAPTER VIII

1. Colenbrander, MC: Calculation of the power of an iris-clip lens for distance vision. Br. J. Ophthal. 57:735-740, 1973

2. Binkhorst, RD: The optical design of intraocular lens implants.Ophthal Surg. 6:17-31, 1975

3. Shammas, HJF: Post-operative anterior chamber depth for anterior chamber lenses. A.I.O.I.S. Journal. 6:153-155, 1980

4. Binkhorst, RD: Intraocular lens power calculation manual: a guide to the author's TI-58/59 IOL power module, ed. 2, New York, 1981,R.D. Binkhorst.

5. Shammas, HJ: Atlas of Ophthalmic Ultrasonography and Biometry. pp. 273-308. The C.V. Mosby Co., St. Louis, 1984.

6. Sanders, DR, Retzlaff, J, Kraff, MC: Comparison of the SRK II formula and the other second generation formulas. J. Cataract Refract Surg. 14:136-141, 1988.

7. Retzlaff, JA, Sanders, DR, Kraff, M: Lens implant power calculation. A manual for ophthalmologists and biometrists. Third Edition. pp. 12-16, 1990. Slack, Inc. Thorofare, N.J.

8. Holladay, JT, Prager, TC, Ruiz, RS, Lewis, JW: Improving the predictability of intraocular lens power calculations. Arch Ophthal 104:539-541, 1986.

9. Holladay, JT, Prager, TC, Chandler, TY, Musgrove, KH: A three-part system for refining intraocular lens power calculations. J. Cataract Refract Surg. 14:17-24, 1988.

CHAPTER IX

SELECTING THE PROPER IOL

The selection of the proper IOL power prior to surgery is an important process and should not be rushed. It is best performed a few days prior to surgery and not at the last minute in the operating room.

Careful discussion with the patient will enlighten the surgeon as to the patient's expectations. Many patients know exactly what they want. On the other hand, there are some patients who will leave the decision up to their ophthalmologist and it becomes incumbent upon the surgeon to make the proper decision.

Some ophthalmologists advocate emmetropia for all their patients while others routinely advocate slight myopia(1-10). Studies have shown that a small amount of myopic astigmatism can enhance the depth of focus of the pseudophakic eye, with adequate 20/30 visual acuity for both near and distance fixation, thus providing spectacle independence. However, not all eyes are the same and certainly not all patients are. The surgeon has to review the patient's needs and expectations, if the eye to be operated is the dominant eye or not, the status of the fellow eye, the IOL power for emmetropia, isometropia and iseikonia and then choose the IOL power accordingly.

1. PATIENT'S NEEDS AND EXPECTATIONS

Emmetropia and ametropia

Most formulas automatically measure the IOL power needed for emmetropia that will allow clear distant vision. Also most calculators and computer programs will give a range of expected postoperative refractive errors with the use of an IOL of a known power; this is helpful if some ametropia is desired. If such a program is not available, the IOL power for ametropia can be calculated by adding the desired refractive error at the cornea to the keratometric readings and using the same formula. For example, a patient with an axial length of 23.5 mm and K readings of 43.5 diopters requires a 20.0 diopter posterior chamber lens for emmetropia. If a postoperative refraction of -0.50 diopters is desired, then the K readings are adjusted to 43.50 -0.50 = 43.0 diopters, that will require an IOL power of 20.7 diopters.

Isometropia

Isometropia is equality of refractive error. The isometropic IOL power is calculated with the regular formula used for emmetropia with one modification: The contact lens refractive power of the fellow eye is added to the K readings of the eye to be operated upon.

For example, if the fellow eye has a -2.00 diopters of myopia at the cornea and the surgeon decides to make the operated eye, that has average K readings of 44.00 diopters, isometric with the fellow eye, then enter into the formula the adjusted K readings of 44.00 - 2.00 =42.00 diopters. The eyes tolerate a 1.5 to 2.0 D difference in refractive error between the two eyes, with no risks of asthenopia and/or diplopia.

Iseikonia

Iseikonia is the equality of retinal image size(11-14). The two eyes have to have the same posterior focal length. In emmetropic eyes, changing the refraction of a pseudophakic eye a certain amount to the myopic side usually results in iseikonia. In eyes with large refractive errors, the problem becomes more complicated and specific formulas are needed for proper calculations (Chapter III). The majority of the patients can ignore aniseikonia of up to +/- 5%, which reflects a refractive error variation of +/- 2.75 D.

The iseikonic IOL power is calculated and seriously considered when the fellow eye is normal with high myopia or high hypermetropia, good vision and does not require surgery. The importance of the isometric and iseikonic IOL power calculations decrease tremendously if the fellow eye has a cataract and surgery is contemplated in the near future.

Accommodating the patient

Patient's expectations usually coincide with their needs and they should certainly be accommodated accordingly (15-19).

- Active patients normally prefer emmetropia, while sedentary patients might prefer slight myopia.
- A patient with hyperopia will most certainly enjoy emmetropia while a patient with high myopia might prefer to remain slightly myopic.

A patient's expectations can differ occasionally from his or her needs due to a possible anisometropia or aniseikonia. This will happen if the fellow eye has normal vision with a certain degree of myopia or hyperopia. The situation is thoroughly discussed with the patient until an acceptable solution is reached. Some interesting cases are presented in this chapter, with alternate solutions. All cases presented in the following sections are from the author's own practice. The calculations are made for a biconvex posterior chamber lens, with the Shammas formula as performed by the author. In each case, the IOL power for emmetropia and isometropia is calculated. The IOL power for iseikonia is only measured in cases of monocular cataracts, with the resultant refractive error and its 5% tolerance range.

The selection of the IOL power and the expected postoperative refractive error should be discussed with the patient prior to surgery.

2. IOL POWER SELECTION IN EMMETROPES

IOL power selection in emmetropes is relatively easy and non-controversial in most cases. For the purpose of discussion, this section includes patients with a refraction ranging between +1.00 and -1.00 D. Young active emmetropes want to remain so, while older sedentary patients might enjoy slight myopia.

The surgeon has to remember that these emmetropes do not tolerate any error in IOL power calculations; they might reluctantly accept induced myopia, but they will rarely accept induced hyperopia.

Monocular cataract in an active patient

Case presentation:

A 68-year-old active patient has a cataract in the right eye. The readings are as follows:

	OD	OS
Vision	20/80	20/25
Refraction	+1.00-1.00 x 90	+0.50 Sph
K readings	44.00/45.00 D.	44.25/44.50 D.
Axial length	22.97 mm	23.02 mm
IOL for emmetropia	20.45 D	20.45 D
IOL for isometropia	19.77 D	19.77 D
IOL for iseikonia	21.94 D	—
Iseikonic refraction	-1.50 sph	—

5% iseikonic tolerance range
-4.25 to +1.25 sph

The choices for the cataractous right eye are:

a. To aim for emmetropia. Most patients enjoy emmetropia. Also in this case, the power for emmetropia is a compromise between the powers for isometropia and iseikonia.

b. To aim for slight myopia by using a higher power IOL. The patient might tolerate a -0.50 sph, but more myopia carries the risk of anisometropia.

Bilateral cataracts in a sedentary patient

Case presentation:
An 82-year-old sedentary patient is referred for cataract surgery to both eyes. The readings are as follows:

	OD	OS
Vision	Hand motion	Count Fingers
Refraction	Plano	Plano
K readings	43.00/43.00 D.	43.00/43.25 D.
Axial length	23.69 mm	23.71 mm
IOL for emmetropia	20.04 D	19.82 D
IOL for isometropia	20.04 D	19.82 D

The IOL power for iseikonia is not required since both eyes have cataracts and require surgery.

The dominant right eye is operated upon first. The choices are:

a. To aim for -0.25 to -0.50 Sph. The patient does not work and does not drive. She spends her days around the house and watching T.V. In aiming for slight myopia, the lady can attend to her needs with no glasses.

b. To aim for emmetropia. This choice is also very acceptable and would have been the choice selection if the patient is still active.

c. To aim for a -2.50 Sph. The induced myopia will allow good reading potential, but the patient will require glasses for distant vision. Furthermore, it is not advisable to induce so much myopia in the dominant eye.

Three months later, the operated right eye has a vision of 20/40 with a refraction of -0.25 -0.25 X 180 deg. The left eye is now scheduled for surgery. The choices are:

a. To aim for -0.50 to -1.00 Sph. The slightly higher myopia in the non-dominant eye can be helpful in allowing the patient to perform all her chores with no glasses. Glasses will only be needed for reading fine print.

b. To aim for emmetropia. Aiming for emmetropia is not a perfect choice for this inactive patient, but the patient will certainly tolerate it. Reading glasses are definitely needed.

c. To aim for -2.50 Sph..A -2.50 D of myopia will give the patient good reading vision, but it will cause a certain degree of anisometropia.

Low power IOL for an emmetrope

In an emmetropic eye, the surgeon expects to use an IOL with a power ranging from 18 to 21 diopters. Occasionally IOL power calculations will call for a lower power than expected.

Case presentation:
A 72-year-old immigrant lady has been complaining of decreased vision in both eyes, of recent onset. Prior to that, she has never worn glasses and has always claimed adequate vision. On examination she has advanced cataracts in both eyes and the following readings:

	OD	OS
Vision	Count fingers	Count fingers
Refraction	No reflex	No reflex
K readings	44.50/44.75 D.	44.50/45.00 D.
Axial length	25.12 mm	25.10 mm
IOL for emmetropia	13.47 D	13.42 D

The case is a real puzzle since the lady never wore glasses. It will be unusual for her to be a real emmetrope, based on the axial length measurements and IOL power calculations. All measurements were repeated and confirmed.

A careful questioning of the patient and of her family revealed that she is illiterate and that she had never visited an ophthalmologist in her native country. One can then postulate that this patient was, indeed, a myope all her life and never knew it!

The choices in this case are:

a. To aim for -0.50 to -1.00 Sph in the right eye and -2.00 to -2.50 in the left eye. This option will give the patient adequate vision for her work at home and her needlepoint work.

b. To aim for emmetropia in both eyes. Although the patient will enjoy uncorrected distance vision, she might be disappointed with the use of reading glasses, since she probably could adequately see for near without glasses.

c. To aim for -2.50 Sph in both eyes. Although the patient will enjoy uncorrected reading vision the distance vision will remain relatively blurry. It is unlikely this patient will be convinced to wear glasses, since she has never worn any before.

High power IOL for an emmetrope

Nuclear sclerosis is well known to induce a myopic shift, and it is not unusual for a hyperopic eye to become emmetropic with the progression of the cataract.

Case presentation:
A 78-year-old patient has bilateral cataracts. The patient has been aware of the presence of cataracts for the past six years, but the vision started deteriorating within the past few months only. The readings area as follows:

	OD	OS
Vision	20/70	20/100
Refraction	+0.75 Sph	+0.50 Sph
K readings	42.25/42.75 D.	42.00/42.50 D.
Axial length	22.21 mm	22.22 mm
IOL for emmetropia	25.85 D	26.15 D
IOL for isometropia	25.19 D	24.49 D

The IOL power for iseikonia is irrelevant in this case due to the decreased vision in the right eye, which will ultimately need cataract surgery too. It is apparent in this case that the patient was a hyperope prior to the development of the cataract.

The choices in this case are indeed limited:

a. To aim for emmetropia or slight myopia (-0.50 to -0.75 Sph). However, the surgeon should remember the high margin of error in these short eyes.

b. To aim for the original hyperopic refraction. This solution is really not desirable.

3. IOL POWER SELECTION IN MYOPES (-1.00 to -3.00 Sph)

Myopes with a refraction ranging between -1.00 and -3.00 Sph usually use their prescription for distance vision and remove their glasses for reading, contrary to higher myopes and to hyperopes who are used to wearing their glasses for distance and reading.

Old sedentary patients usually prefer to remain near-sighted while young active myopes are more likely to request emmetropia. Surgeons should not aim for "bull's eye" emmetropia, because they can push the patients into hyperopia in case of a minor error in IOL power calculations. Instead they should aim toward a -0.50 to a -1.00 Sph, a refraction that most myopes seem to enjoy the most for distance vision and reading of large print.

Bilateral cataracts in a sedentary patient

Case presentation:

A 70-year-old patient is examined for decreased vision in both eyes due to cataract formation. The patient had a stroke two years earlier and is confined to a wheelchair. The following readings are taken:

	OD	OS
Vision	20/100	20/200
Refraction	-2.25-0.50 X 60	-2.50 Sph
K readings	44.00/44.75 D.	44.50/44.75 D.
Axial length	24.32 mm	24.30 mm
IOL for emmetropia	16.23 D	15.89 D
IOL for isometropia	19.56 D	19.22 D

The IOL for iseikonia is irrelevant in this case because both eyes have cataracts that require surgery.

Upon discussion with the patient, it is understood that the patient is an avid reader and does not use her glasses for that task; she only wears single vision glasses for watching T.V. One more important piece of information: the patient hates bifocals!

The choices in this case are:

a. To aim for -2.50 Sph in OU. A -2.50 D of myopia will keep the status quo for this elderly patient. She will continue reading without glasses and will use her single vision glasses when needed.

b. To aim for -1.00 to -1.50 Sph in the dominant eye and -2.50 Sph in the other eye . This combination will provide functional and reading vision without much anisometropia between the two eyes.

c. To aim for emmetropia in one eye and -2.50 Sph in the other eye. Unfortunately, the difference between the two eyes will cause a certain degree of anisometropia.

d. To aim for emmetropia in both eyes. This choice will be a disservice to this patient who is confined to a wheelchair. Furthermore, in this case, it will be going against the patient's wishes and expectations.

Monocular cataract in an active patient

Case presentation:
A 66-year-old patient has a monocular cataract in his dominant right eye. The readings are:

	OD	OS
Vision	20/100	20/20
Refraction	-2.25 Sph	-2.00 Sph
K readings	43.00/43.00 D.	42.75/43.00 D.
Axial length	24.53 mm	24.45 mm
IOL for emmetropia	17.42 D	17.84 D
IOL for isometropia	20.10 D	20.80 D
IOL for iseikonia	22.33 D	—
Iseikonic refraction	-3.75 Sph	—

5% iseikonic tolerance range
-1.00 to -6.50 Sph

The right eye is scheduled for surgery. The choices in this case are:

a. To aim for -2.25 Sph. A -2.25 D of myopia will keep the status quo, eliminating the need for reading glasses. It does not cause any anisometropia or any significant amount of aniseikonia.

b. To aim for -0.75 Sph. to -1.00 Sph. The -0.75 D of myopia is a compromise between isometropia and emmetropia. It provides this active patient with adequate functional vision, but he will still need glasses for distance vision and possibly, for reading too. Furthermore, it will cause a certain degree of aniseikonia, but in this case it can be tolerated by the patient.

c. To aim for emmetropia. Emmetropia provides this active patient with clear distance vision, but the difference between the two eyes can be intolerable. This choice is viable if the patient insists on emmetropia and if the left eye can be fitted with a contact lens or treated surgically. Occasionally, some patients will accept monocular vision, using the dominant eye for distance and the non-dominant eye for reading.

Monocular cataract in one myopic eye

Case presentation:
A 62-year-old patient has decreased vision in the left eye. The patient could not tolerate any prescription for the condition and has never worn glasses. The readings from her eyes are:

	OD	OS
Vision	20/30	20/100
Refraction	P1 -0.50 X 90	-2.75 Sph
K readings	44.00/44.00 D.	43.50/44.00 D.
Axial length	23.43 mm	24.60 mm
IOL for emmetropia	19.56 D	16.22 D
IOL for isometropia	23.21 D	16.55 D
IOL for iseikonia	—	22.44 D
Iseikonic refraction	—	-4.75 Sph
5% iseikonic tolerance range		-2.00 to -7.50 Sph

The choices in this case are:

a. To aim for -2.50 to -2.75 Sph. The clinical history reveals that this patient has never worn glasses. This can only mean that the patient always had monocular vision, using the right eye for distance and the left eye for reading. A myopic refraction also will prevent any large amount of induced aniseikonia. This choice will keep the status quo and the patient very happy.

b. To aim for emmetropia. Although many surgeons will opt first for this choice, it will be a mistake in this case. A thorough discussion with the patient will immediately clarify the issue. This choice is, of course, the preferred choice if the myopia in the left eye is secondary to the cataract development, and not to a difference in the axial length. This choice also will be preferred if the myopic left eye, with the cataract, is the dominant eye.

4. IOL POWER SELECTION IN HIGH MYOPIA (over - 3.00 D)

Patients with high myopia develop cataracts at a younger age than emmetropes and hyperopes. The surgeon often has to deal with young, active patients and their high expectations.

Bilateral cataracts in an active patient

Case presentation:

This 50-year-old nurse has decreased vision in both eyes due to cataracts. She always has been myopic and could never tolerate contact lens wear. The readings are:

	OD	OS
Vision	20/80	20/70
Refraction	-8.00 -2.50 X 95	-10.25 Sph
K readings	42.00/44.00 D.	43.50/44.00 D.
Axial length	29.30 mm	29.55 mm
IOL for emmetropia	5.65 D	4.15 D
IOL for isometropia	18.78 D	16.10 D

The IOL power for iseikonia is not necessary in this case since both eyes have cataracts and require surgery.

The choices in this case are:

a. To aim for -0.50 Sph in the dominant right eye and for -1.50 Sph to -2.00 Sph in the fellow eye to free the patient from glasses. However, reaching the planned post-operative refraction might be more difficult in long eyes because of the accentuated curvature of the posterior pole. Furthermore, all available formulas are less accurate in long eyes. In aiming for a -0.50 D of myopia, the surgeon keeps a comfortable margin of error between Plano and -1.00 Sph.

b. To aim for emmetropia. It is not unusual for such patients to insist on bilateral emmetropia. This choice would be popular with patients who have been wearing corrective contact lenses for distance vision. The surgeon must remember that some young high myopes have not yet experienced presbyopia and might expect good uncorrected vision for distance and near. Furthermore, in aiming for "bull's eye" emmetropia, the surgeon can induce some degree of hyperopia resulting in a very unhappy patient. A good discussion with the patient is essential in this case.

c. To aim for -2.50 Sph. This choice is more suitable for patients who have always worn glasses for distance vision. These patients will enjoy reading without any correction and will wear their new, thinner glasses for driving or watching TV.

Monocular cataract in an active patient

Case presentation:
This 47-year-old executive has decreased vision in one eye due to a posterior subcapsular cataract. He has always been myopic. He wore contact lenses when he was young, but has not worn any for the past 10 years. The readings are:

	OD	OS
Vision	20/20	20/100
Refraction	-7.00 -0.25 X 90	-8.25 Sph
K readings	43.00/43.25 D.	43.00/43.50 D.
Axial length	25.9 mm	26.1 mm
IOL for emmetropia	13.40 D	12.70 D
IOL for isometropia	24.05 D	21.94 D
IOL for iseikonia		24.25 D
Iseikonic refraction		-9.00 Sph
5% iseikonic tolerance range		-6.25 to -11.75 Sph

The patient prefers emmetropia but leaves the choice to the surgeon. A good discussion in this difficult case is essential. The patient can be offered the following choices:

a. To aim for emmetropia or slight myopia. It is the dream of most young myopes to have clear uncorrected vision and this is definitely a golden opportunity for this young executive. Please note the difference between the IOL power for emmetropia and the IOL power for iseikonia in this case. Aiming for emmetropia will most certainly create anisometropia and aniseikonia. This can be avoided by decreasing the myopia in the right eye either with a contact lens (the patient was a contact lens wearer) or with corneal surgery. More daring surgeons will even propose a clear lensectomy to the right eye.

b. To aim for -6.50 Sph. Leaving the operated eye with this myopia will avoid a significant degree of anisometropia and aniseikonia. However, the patient is missing his chance to get rid of his glasses. Some surgeons can make a case for an extracapsular cataract surgery and the insertion of a high power anterior chamber lens to cause -6.50 of myopia. It is then replaced with a permanent, weaker posterior chamber IOL when the fellow eye develops a cataract.

Monocular cataract in unilateral axial myopia

Case presentation:
A 42-year-old patient has been complaining of decreased vision in the left eye, due to a dense cataract. The readings are as follows:

	OD	OS
Vision	20/20	Count fingers
Refraction	-0.50 Sph	-10.50 Sph
K readings	43.50/43.50 D.	43.00/43.00 D.
Axial length	23.48 mm	29.7 mm
IOL for emmetropia	20.06 D	4.85 D
IOL for isometropia	33.55 D	5.51 D
IOL for iseikonia	—	23.64 D
Iseikonic refraction	—	-14.50 Sph
5% iseikonic tolerance range		-11.75 to -17.25 Sph

It is not unusual for the myopia in the non-dominant eye to go unnoticed in cases of Unilateral Axial Myopia due to the opaque media. An uninformed technician can disregard the axial length measurements, blame it on a difficulty with the A-scan echograms and record a reading similar to the normal fellow eye, resulting in a very large postoperative error in refraction.

The question in our case is whether the myopia in the left eye is secondary to the cataract formation or whether it is due to an axial myopia. Axial length measurements show a difference of 6.22 mm between the two eyes.

The choices in this case are as follows:

a. To aim for emmetropia or a slight myopia. Emmetropia or slight myopia will restore isometropia and partial vision; it also increases the field of vision. Emmetropia or slight myopia will not restore iseikonia, but this does not seem to be a problem because of the associated amblyopia.

b. To aim for -2.50 Sph. The patient uses monocular vision. If the left eye recovers adequate vision, it will improve the reading vision without wearing glasses. Unfortunately, the eye is usually amblyopic and vision rarely improves more than 20/60.

c. To aim for -7.00 or -8.00 Sph. Leaving the eye highly myopic is not a viable alternative in this young patient.

5. IOL POWER SELECTION IN HYPEROPIA

Hyperopic patients look forward to emmetropia after the cataract surgery. Older sedentary patients might even enjoy a certain degree of myopia.

Bilateral cataracts, moderate hyperopia

Case presentation:
A 71-year-old patient has bilateral cataracts and early macular degeneration. The readings from the eyes are:

	OD	OS
Vision	20/60	20/200
Refraction	+3.75-1.50 X 100	+3.50-0.75 X 85
K readings	42.50/43.50 D.	42.50/43.00 D.
Axial length	22.72 mm	22.70 mm
IOL for emmetropia	23.33 D	23.73 D
IOL for isometropia	19.13 D	19.71 D

The IOL for iseikonia is irrelevant in this case because both eyes have cataracts and will ultimately need surgery.

The choices in this case are:

a. To aim for emmetropia in both eyes. The presence of mild astigmatism will also provide a certain depth of focus. The patient will need glasses only for reading.

b. To aim for +1.50 Sph in the left eye. This option is to be considered only if the left eye is to be operated on, leaving the right eye with a 20/60 vision. Occasionally, an older patient is reluctant to have both eyes operated on, especially if the anticipation of vision improvement is not certain due to the concomitant presence of macular degeneration.

c. To aim for -2.50 Sph in both eyes. This option is not very desirable and is to be considered only if the patient requests it. The patient will have to understand that distance glasses are required and that there will be an image minification due to the induced myopia.

Monocular cataract, high hyperopia

Case presentation:

A 66-year-old patient suffers from decreased vision in his dominant right eye due to cataract. The left eye has minimal cataract changes. The readings are:

	OD	OS
Vision	20/80	20/30
Refraction	+6.75-0.50 X 7	+6.00 Sph
K readings	44.75/45.50 D.	45.00/45.25 D.
Axial length	20.91 mm	20.95 mm
IOL for emmetropia	27.54 D	27.37 D
IOL for isometropia	19.25 D	18.37 D
IOL for iseikonia	21.47 D	----
Iseikonic refraction	+4.25 Sph	----

5% iseikonic tolerance range
+1.50 to +7.00 Sph

This is an unusual case that requires some discussion with the patient as to the best modality of treatment. The choices in this case are as follows:

a. To aim for emmetropia. This 66-year-old active patient will certainly enjoy good uncorrected vision in his dominant eye. However, emmetropia will cause an anisometropia and aniseikonia that have to be addressed and discussed with the patient. To avoid these problems, the left eye can be fitted with a contact lens, provided that the patient can tolerate it. However, hyperopes are not as enthusiastic about contact lens wear as myopes are. Another possibility is for the patient to ignore the relatively blurred vision in the left eye and depend on the right eye for vision. The left eye is starting a cataract and will have to be operated on sooner or later; the patient might opt for "sooner" than for "later" after he discovers clear uncorrected vision in the right eye.

b. To aim for +3.00 Sph to +4.00 Sph. The patient can tolerate 2.00 to 2.50 D of anisometropia without causing any aniseikonia. This solution can be considered if the patient wants to keep on wearing the glasses and does not want to wear a contact lens or have surgery to the left eye.

Monocular aphakia

Case presentation:
A 75-year-old lady has been aphakic in her right eye for the past ten years and has been fitted successfully with a contact lens, until recently. The left eye is pseudophakic with a 21.0 diopter posterior chamber lens in place. The readings are as follows:

	OD	OS
Vision	20/50	20/30
Refraction	+13.75-1.50 X 115	+0.50 Sph
K readings	43.50/44.50 D	44.00/44.00 D.
Axial length	22.44 mm	22.50 mm
IOL for emmetropia	23.00 D	22.78 D
IOL for isometropia	22.33 D	—
IOL for iseikonia	22.12 D	—
Iseikonic refraction	+0.67 Sph	—

5% iseikonic tolerance range
-2.08 to +3.42 Sph

The choices for the right eye:

a. To aim for emmetropia. In this case, emmetropia will restore isometropia and iseikonia. It seems to be the ideal choice.

b. To aim for -0.75 Sph. Although a -0.75 D of myopia gives the patient some near vision capability, it will create a 1.25 diopter difference between the two eyes.

6. IOL POWER SELECTION IN CHILDREN

At birth, the child's eye measures approximately 15 mm and increases rapidly to reach 21 mm by the age of two years. The growth continues afterward at a much slower rate to average 23.5 mm in adulthood. Similarly, the cornea flattens from an average of 51 diopters to an average of 44 diopters between birth and the age of two years; the cornea flattens an additional 0.50 diopter to adulthood(20).

If surgery is done within the first two years of life, and an implant is inserted, a large myopic shift is to be expected a few years later. This has led many surgeons to remove the cataract and fit the eye with a contact lens instead of inserting an IOL.

When surgery is performed after the age of two years, the expected myopic shift ranges from 4 to 6 diopters as the axial length increases with time. Many surgeons recommend undercorrecting the IOL power by around 3 diopters to partially compensate for the myopic shift; any greater undercorrection would lead to anisometropia and would conflict with amblyopia treatment. The residual myopia in adulthood can be treated with spectacles, contact lenses or corneal surgery(21,22).

Axial length measurement might be a challenge in these young patients. If this is the case, the corneal power and the axial length should be measured under general anesthesia.

REFERENCES - CHAPTER IX

1. Binkhorst, RD: Biometric A-Scan ultrasonography and intraocular lens power calculation, In Current concepts in cataract surgery: selected proceedings of the Fifth Biennial Cataract Surgical Congress, Emery, J.M. (ed.). pp. 175-182, St. Louis, 1978,The C.V. Mosby Co.

2. Boerner, CF, Thrasher, BH: Results of monovision correction in bilateral pseudophakes. AIOIS J. 10:49-50, 1984.

3. Drews, RC: A practical approach to lens implant power. AIOIS J. 3:170-176, 1977.

4. Drews, RC: The determination of lens implant power. Ophthalmic Surgery. 20:625-637, 1989.

5. Fechner, PU, Kania, J, and Kienzle, S: The value of a zero power intraocular lens. J. Cataract and Refractive Surgery. 14:436-440, 1988.

6. Gill, JP: Minimizing postoperative refractive error. Cont. Intraocul. Lens. Med. J. 6:56-59, 1980.

7. Hoffer, KJ: Preoperative cataract evaluation: intraocular lens power calculation. Int. Ophthalmol. Clin. 22:37-75, 1982.

8. Menezo, JL, Cisneros, A, and Harto, M: Extracapsular cataract extraction and implantation of a low power lens for high myopia. J. Cataract and Refractive Surgery. 14:409-412, 1988.

9. Shammas, HJ: Spectacle correction desired after cataract removal. J. Cat. Refr. Surgery. 17:101-102, 1991.

10. Kora, Y, Yagushi, S, Inatomi, M. and Ozawa, T: Preferred postoperative refraction after cataract surgery for high myopia. J. Cat. Refr. Surg. 21:35-38, 1995.

11. Huber, C, and Binkhorst, C: Iseikonic lens implantation in anisometropia. AIOIS J. 5:194-198, 1979.

12. Miyaki, S, Awaya, S, and Miyaki, K: Aniseikonia in patients with a unilateral artificial lens, measured with Aulhorn's phase difference haploscope. AIOIS J. 7:36, 1981.

13. Troutman, RC: Artiphakia and aniseikonia. Am. J. Ophthalmol. 56:602-639, 1963.

14. van der Heijde, GL: The optical correction of unilateral aphakia.Trans. Am. Acad. Ophthal. Otolaryngol. 81:80-88, 1976.

15. Binkhorst, RD: Pitfalls in the determination of intraocular lens power without ultrasound. Ophthal. Surg. 76:69-82, 1976.

16. Holladay, JT, Rubin, ML: Avoiding refractive problems in cataract surgery. Survey Ophthalmology. 32:357-360, 1988.

17. Hoffer, KJ: Biometry of 7500 cataractous eyes. Am. J. Ophthalmol. 90:360-368, 1980.

18. Kratz, RP, Shammas, HJ: Color Atlas of Ophthalmic Surgery. Cataracts. pp. 13-19. J.B. Lippincott Co., Philadelphia. 1991.

19. Shammas, HJ: Atlas of Ophthalmic Ultrasonography and Biometry. PP.273-308. The C.V. Mosby Co., St. Louis, 1984.

20. Gordon, RA and Donzis, PB: Refractive development of the human eye. Arch. Ophthalmol. 103:785-789, 1985.

21. Sinskey, RM and Patel, J: Posterior chamber intraocular lens implants in children: report of a series. Am Intra-Ocular Implant Soc. J. 9:157-160, 1983.

22. Vasavada, A and Chauhan, H: Intraocular lens implantation in infants with congenital cataracts. J. Cat. Refract. Surg. 20:592- 598, 1994.

CHAPTER X

ERRORS IN IOL POWER CALCULATIONS AND THE UNEXPECTED POSTOPERATIVE RESULT

With the refinement of IOL power formulas, patients and surgeons are now expecting a more accurate final refraction in the operated eye. Twenty years ago, results within +/- 3.0 D. were acceptable and deemed superior to any aphakic correction(1-4). Nowadays, our results are accurate to within +/- 0.5 D. in most cases.

All available IOL power formulas are based on mathematical equations with modifications obtained from clinical results. They all require three entries that are the axial length, the corneal power and a constant (A,S or ACD). An error in the determination of any of these entries will undoubtedly result in an unexpected postoperative result.

Occasionally, the error is very large and the IOL has to be exchanged. In most cases, the error ranges between 2 and 5 diopters, resulting in an induced myopia or hyperopia in the operated eye. The patient's reaction varies according to:

- His or her expectations
- Which eye is involved: dominant v/s non-dominant
- The amount of anisometropia and aniseikonia.

In myopes and hyperopes who have always worn glasses for distance vision, an error in IOL power calculations is covered up by a postoperative change in the lens prescription and does not always result in a postoperative disappointment to the patient. Emmetropes, on the other hand, have higher expectations; an induced error in the dominant eye with anisometropia and/or aniseikonia will certainly result in patient disappointment. In such cases, the surgeon and the patient have multiple choices:

- Fit the patient with new glasses
- Fit the operated eye with a contact lens
- Fit the non-operated eye with a contact lens
- Corneal surgery, i.e. radial keratotomy, laser photorefractive keratectomy
- Exchange the implant.

These choices are discussed with the patient and the appropriate course of action is taken.

1. THE LARGE ERROR

In the early 1970s, the axial length was not routinely measured prior to surgery and a clinical method was used to calculate the IOL power:

$$P = 18 + 1.5\,R$$

where P is the power of the iris-supported IOL needed for emmetropia, and R is the eye's refractive error prior to the development of the cataract.

The formula was accurate to within +/- 3 D. Occasionally, large errors occurred after surgery. A classical example would be the case of the younger patient with a mature cataract in one eye and a normal emmetropic fellow eye. The cataract is extracted and an 18 diopter iris-supported lens is implanted. After the surgery, the refraction is a totally unexpected "-9.00 diopters". This large error was labeled the "9 Diopter Surprise" and was correctly attributed to an unsuspected and undetected longer axial length in the operated eye.

Over 20 years have passed and surgeons now routinely measure the axial length and the corneal power, and use sophisticated IOL power formulas. Unfortunately, large errors are still being reported.

A large postoperative refractive error is often due to a large error in axial length measurement occurring in a very long or very short eye(5-8); in certain cases, it is associated with the use of an original formula such as the SRK-I formula and/or a misjudgment on the part of the surgeon. Large errors have also been reported with the use of mislabeled intraocular lenses.

Large error in Unilateral Axial Myopia

Case presentation:
A 52-year-old lady was seen in consultation after surgery to her right eye. On examination, the visual acuity was 20/80 in the right eye with a -9.50 D correction and 20/25 in the left eye with a +0.25 D correction. Prior to the cataract surgery, both eyes were measured in the surgeon's office using a contact technique, and reported to be 23.4 mm in both eyes. The technician also noted more difficulty in the measurement of the right eye. A repeat measurement of both eyes revealed an axial length of 31.2 mm in OD and 23.62 mm in OS using an immersion technique. Upon advice, the surgeon removed the implant without replacing it. Three months after the second surgery the visual acuity was 20/70 with a refraction of -1.00 D.

In all such cases where the fundus is obscured by a mature cataract, a B-scan examination rules out any vitreoretinal pathology. In a case of Unilateral Axial Myopia, it will show the elongation of the eye with an increase in the retinal curvature that explains the technician's difficulty in obtaining an accurate retinal spike. Instead, the technician must have obtained an oblique measurement or, out of frustration, reported an equal measurement to the normal fellow eye. There were no ultrasound tracings available for review in this case.

Unilateral Axial Myopia(8) is a rare congenital anomaly, causing anisometropia and amblyopia in the affected eye. It remains undetected throughout childhood; patients go to the ophthalmologist when they lose vision in the affected eye because of the cataract's progression. When the patient is examined for the first time, the myopia is not detected for two reasons: (1) the patient is not wearing the myopic correction because of the associated anisometropia and amblyopia, and (2) the preoperative refraction is unreliable because of the opaque media.

Unilateral Axial Myopia is a developmental anomaly causing unilateral visual deprivation during the critical first year of life. This will result in early amblyopia with a visual acuity of around 20/200. Surgery is recommended when the cataract becomes mature and visual acuity drops to light perception. It restores isometropia and partial vision, and increases the field of vision. One should not attempt to restore iseikonia. In these long eyes, iseikonia would have required a stronger intraocular lens to be inserted during surgery, resulting in a high myopic correction postoperatively. The induced aniseikonia is not a problem because of the associated amblyopia.

Large error in the short and microphthalmic eye

Case presentation:

This 58-year-old man has been aware of his eye problems since early childhood. Through the years, he has been extensively evaluated at major university centers and was diagnosed with bilateral microphthalmia, high hyperopia and a delineated form of retinitis pigmentosa. Six weeks earlier, he underwent a phacoemulsification cataract extraction with primary implantation of a 21.5 diopter posterior chamber lens. Upon the patient's request, the surgeon aimed towards emmetropia, but ended with a post-operative refraction of +10.75 +0.50 X 180, improving his vision to 20/50. His visual acuity was 20/100 in the left eye with +10.50 +0.75 X 15. The corneal diameter was 9.5 mm in both eyes. He was referred to our office for repeat axial length measurement and IOL power calculations. Bilateral axial length measurements were performed with the Kretz 7200 MA ultrasound unit using an immersion technique. Both eyes measured 16.4 mm with keratometric readings of 51 diopters. The anterior chamber depth measured 2.9 mm, the lens thickness 4.0 mm and the vitreous cavity 9.5 mm. The Shammas formula was used. Both eyes were found to need 44 diopter lenses for emmetropia. Due to the microphthalmia and the unusual hyperopia, a conservative use of 38 diopter lenses was recommended.The ophthalmologist operated first on the left eye, removing the cataract and inserting a 38 diopter posterior chamber lens. Two months later, the wrong implant was removed from the right eye and replaced with a 38 diopter lens. The final refraction was +1.50 Sph. in both eyes.

Intraocular lens power calculations in microphthalmic eyes(7) present special problems due to the extreme shortening of the axial length and the lack of data in such cases. In this patient, a 38 diopter lens was used, yielding a refraction of +1.50 diopters. The exact IOL power needed for emmetropia was recalculated using the available postoperative data and was found to be 41.5 diopters.

The IOL power calculations performed in the referring ophthalmologist's office resulted in an 11 diopter postoperative refraction surprise after the first operation. A 21.5 diopter lens was used instead of a 41.5 diopter. The 20 diopter lens power error was due to a wrong axial length measurement by the referring ophthalmic surgeon and the use of the SRK-I formula:

1. The technician used a contact technique to measure the axial length and had problems identifying the echospikes. He reported a measurement of 20.0 mm instead of 16.4 mm. Since no representative pictures were taken, one can only postulate that the measurement was taken between the corneal and some scleral or orbital spike instead of the retinal spike.

2. The technician used the SRK-I formula to calculate the IOL power. The use of the SRK-I formula with the 16.4 mm axial length calls for a much weaker implant. The two errors resulted in a total of 20 diopters. Furthermore, the technician and the surgeon should have realized that a 21.5 D. lens can never suit a long-standing hyperope with a spherical equivalence of +11.00 Sph., and should have questioned the IOL calculations prior to the planned surgery.

High-power implants are not always available for these short eyes. Some surgeons are now attempting to correct the hyperopia by inserting two implants in the same eye after removing the cataract. The first implant is inserted in the capsular bag, and the second implant is placed in front of it, either in the sulcus or in the anterior chamber. Complex IOL power calculations are required to reach emmetropia.

Large error due to misjudgment

Case presentation:
This 69-year-old patient has been emmetropic all his life and has worn glasses only for reading. After noticing a decrease in vision in his right eye he consulted his ophthalmologist who recommended surgery for the developing cataract. Measurements obtained from the operating surgeon revealed the following preoperative values:

	OD	OS
Vision	20/60	20/30
Refraction	Plano-0.50 X 90	Plano-0.25 X 85
Axial length	20.41 mm	20.45 mm
K readings	43.50/43.75 D.	43.50/43.50 D.

The IOL calculations called for a 30 D. implant for emmetropia. The surgeon removed the cataract in the right eye through a small incision and inserted a 30 diopter implant in the capsular bag. The next week, the right eye improved to 20/40 with a refraction of -9.50 Sph.

After investigation, it was noted that the biometer was sent for repairs and that the technician had used a loaner on this specific day, and that the corneal gate had been moved to the right thus erroneously measuring the axial length from the lens surface to the retina instead of the cornea to the retina.

A remeasurement of both eyes with the regular biometer revealed an axial length of 23.5 mm in both eyes, necessitating the use of a 19.0 diopter IOL for emmetropia. The patient was referred to an outside consultant for a lens exchange.

The error in axial length measurement in this case had been compounded by an error in judgment on the part of the surgeon. The patient had a long history of emmetropia with good refraction and adequate vision.The surgeon should have realized that such a patient can never require a high power implant. A repeat axial length measurement either in the presence of the surgeon or by an outside consultant is recommended in such cases to avoid major postoperative problems.

Large error due to a mislabeled IOL

In the early days of lens implantation, surgeons encountered some cases of mislabeled IOL due to the fact that some IOL companies did not use adequate quality control during the manufacturing process. Unfortunately, such cases are still encountered in Europe, the Third World countries, and even in the United States.

Case presentation
This 62-year-old myopic man underwent an uneventful extracapsular cataract extraction with primary insertion of a 13.0 diopter lens to achieve slight myopia in his dominant eye. The left eye was already pseudophakic. The preoperative measurements were as follows:

	OD	OS
Vision	20/100	20/20
Refraction	-9.50 Sph	-1.00-0.50 X 172
Axial length	26.25 mm	26.10 mm
K readings	44.00/44.00 D	44.50/45.00 D

Postoperatively, the eye healed uneventfully, but the patient kept complaining that he could not see without glasses and that the vision was very clear with his old prescription. The vision in the right eye was 20/30 with a refraction of a -9.00 Sph.

The patient was referred for consultation. The axial length measurement was correct and no staphylomas were detected on B-scan examination. The K readings were steeper by only 0.25 D and the implant was in good position.

Upon advice, the surgeon removed the implant and measured it in the operating room to be 23 diopters. It was replaced by a 13.0 diopter implant and the patient later achieved 20/25 vision with a correction of -0.50 Sph.

2. INDUCED MYOPIA

When confronted with an unexpected postoperative myopia, the following steps are followed in search of the cause(9-21).

- Both eyes are remeasured to rule out preoperative shorter axial length measurement, which is the most common cause of error in the presence of an induced myopia.
- The corneal power is recalculated to rule out a postoperative steepening of the cornea.
- The implant's position is reassessed to rule out a forward placement with or without tilting of the IOL within the eye.
- In long and short eyes, the IOL formula is rechecked for accuracy.

Induced myopia due to shorter axial length measurement

A shorter axial length can occasionally be measured prior to cataract surgery. The causes of such a shorter measurement and how to avoid them are detailed in Chapter VI. The use of a shorter axial length measurement in the IOL power calculations will call for the use of a stronger IOL than is required, resulting in an induced myopia in the final postoperative refraction.

Case presentation:
A 64-year-old male was seen in consultation three months after cataract surgery to the left eye, because of an unexpected induced myopia. The preoperative records from his surgeon's office revealed the following parameters:

	OD	OS
Vision	20/25	20/50
Refraction	Plano -0.50 X 82	Plano -0.75 X 103
Axial length	not taken	24.2 mm
K readings	not taken	41.50/42.50 D

The patient had undergone an uneventful planned extracapsular cataract surgery with primary implantation of a 19.50 D. planoconvex posterior chamber lens. Postoperatively, the uncorrected visual acuity was very blurry for distance. A refraction revealed an induced myopia in the operated eye. The patient was referred for consultation and recalculation of the IOL power. On examination the following parameters were noted:

	OD	OS
Vision	20/25	20/25
Refraction	Plano -0.25 X81	-3.00 -0.75 X 103
Axial length	25.4 mm	25.5 mm
K readings	42.00/42.50 D	42.00/42.25 D

Power for emmetropia using a planoconvex lens

Shammas	15.26 D	15.16 D
SRK-T	14.97 D	14.84 D
Holladay	15.28 D	15.12 D

The 1.3 mm shorter axial length measurement prior to the cataract surgery accounted for the induced myopia in this case. The patient could not tolerate the induced anisometropia. The patient was referred back to his surgeon for a lens exchange.

Case presentation:

This 60-year-old man was seen in consultation for an unexpected postoperative induced myopia in his left eye. The right eye was already pseudophakic.

Preoperative records from his surgeon's office revealed the following parameters:

	OD	OS
Vision	20/20 pseudophakic	20/50
Refraction	-0.25 -0.25 X 180	-5.25 -0.50 X 170
Axial length	26.65 mm	26.25 mm
K readings	42.00/42.25 D	42.00/42.50 D

Aiming for slight myopia, the surgeon used a 14.0 diopter planoconvex posterior chamber lens. After the surgery, the left eye was more near sighted than expected.

During consultation, the following parameters were recorded:

	OD	OS
Vision	20/20	20/20
Refraction	-0.25 -0.25 X 180	-2.75 Sph
Axial length	26.75 mm	27.22 mm
K readings	42.00/42.25 D	42.25/42.50 D

Power for emmetropia using a plano-convex IOL:

Shammas	12.05 D	10.63 D
SRK-T	11.64 D	10.21 D
Holladay	11.48 D	9.91 D

There was an error of almost 1 mm between the measurement taken during consultation and the preoperative measurements taken in the surgeon's office. This error accounted for the induced myopia in this patient's left eye.

The patient was disappointed because he was promised near emmetropia. The different alternatives were explained to him. They included:

- Wearing a contact lens to the left eye and using reading glasses for close-up vision.

- Undergoing radial keratotomy to correct the myopia.

- Exchanging the implant with a weaker one, at which time he will still need reading glasses.

- Leaving it the way it is. The left eye is, fortunately, his non-dominant eye and he can learn to use monocular vision with the right eye focused for distance vision and the left eye for reading.

The third choice was the most logical for this active 60-year-old man. Reexamination, three months later revealed he had adjusted to monocular vision and was wearing his distance correction for driving only.

Induced myopia due to corneal steepening

It is extremely unusual for the cornea to steepen after a routine phacoemulsification cataract extraction. Occasionally, tight sutures induce with-the-rule astigmatism, with corneal steepening in the vertical axis; suture removal will restore the preoperative corneal curvature.

Induced myopia due to corneal steepening is more likely to occur after a triple procedure of corneal transplant-ECCE-lens implantation, or of corneal transplant with IOL removal and replacement.

Case presentation:
A 78-year-old man underwent a corneal transplant with a lens exchange for corneal decompensation in his right eye two years after cataract surgery. The patient's vision in the right eye was finger counting at 2 feet with -0.50 Sph; the axial length measured 24.46 mm with distorted K readings of 42.00 diopters. The left eye was pseudophakic with a vision of 20/30 and a refraction of +0.50 -1.75 X86; the axial length measured 24.35 mm and the K readings 43.50 X 44.00 diopters. Planning for emmetropia, a 17.0 diopter anterior chamber lens was inserted during surgery; at the same time, a 7.0 mm corneal button was excised and replaced with a 7.5 mm transplant. After surgery, the eye healed uneventfully and the vision improved to 20/40 with a refraction of -5.25-0.50 X 60; the cornea measured 47.00 X 48.00 diopters with a 5.5 diopter steepening compared to the preoperative readings.

Induced myopia after a triple procedure can be avoided by following three major guidelines:

- Use a corneal button that is a 0.2 mm larger than the recipient, instead of a 0.5 mm larger button.
- Perform a 7.5 mm and, whenever possible, an 8.0 mm corneal transplant instead of a 7.0 mm transplant.
- Allow for a 1 to 2 diopter myopic shift if, preoperatively, the cornea is flatter than the cornea of the fellow eye and, if the cornea is scarred with distorted readings, one should use the K readings of the opposite eye.

Induced myopia due to a displaced implant

During ECCE surgery, the posterior chamber implant is usually placed in the capsular bag. However, in certain cases, the implant will have one of its loops in the bag and the second loop in the ciliary sulcus. The implant will tilt inside the eye and sit closer to the cornea than expected, causing an induced myopia. The condition is exaggerated if the anterior chamber becomes shallow in the immediate postoperative period due to a wound leak, or a concomitant filtering surgery.

Case presentation:
This 65-year-old lady underwent a combined trabeculectomy-ECCE-lens implant to the right eye. The preoperative measurements were as follows:

	OD	OS
Vision	20/200	20/40
Refraction	+1.50 Sph	+1.00-0.25 X 180
Axial length	23.21 mm	23.24 mm
K readings	42.00/42.25 D	42.00/42.50 D

Planning for emmetropia, a 21.5 D posterior chamber lens was used. Postoperatively, the visual acuity improved to 20/30 with a refraction of -1.25 Sph. The intraocular pressure was within normal limits with a good filtering bleb superiorly. The anterior chamber was slightly shallow with an atonic pupil. The implant was tilted with its superior edge caught in front of the pupil. The distance between the anterior corneal surface and the implant was measured with the A-scan biometer and found to be 2.8 mm instead of the expected 4.2 mm. The difference in the anterior chamber depth accounted for the induced postoperative myopia.

The patient was very happy with the result, especially with the fact that she now can read without glasses.

Case presentation:
This 80-year-old lady consulted for decreased vision in her left eye after cataract surgery, performed six months earlier. A review of the surgeon's files revealed the following parameters:

	OD	OS
Vision	20/30	20/40
Refraction	+0.50-1.00 X 124	-0.50-0.75 X 119
Axial length	23.42 mm	23.41 mm
K readings	43.50/43.50 D	43.50/44.00 D
IOL for emmetropia	18.88 D	18.63 D

A 19 D. meniscus posterior chamber lens was used after an uneventful phacoemulsification. Four weeks after surgery, the vision was 20/20 with a correction of Plano -1.00 X 87.

At the two month visit, the patient's refraction had changed to -1.50 -0.75 X 120 and the best corrected vision was 20/30. The patient was disappointed with her vision in spite of new glasses.

Six months after the surgery she was complaining that the vision was much blurrier than before the surgery. The following parameters were noted:

	OD	OS
Vision	20/40	20/50
Refraction	+0.25-0.50 X 120	-2.50-1.00 X 90
Axial length	23.43 mm	23.43 mm
K readings	43.50/43.50	43.25/44.25
IOL for emmetropia	19.35 D	19.03 D

There were no major differences in the axial length measurements, K readings and IOL power calculations. However, the left eye's myopia had increased by -2.50 D. On examination, the implant was quite decentered and the superior edge of the optical portion of the implant was clearly seen in the pupillary area and the superior loop was embedded superiorly in the capsular bag. The inferior loop could not be seen, but most probably was located in the ciliary sulcus with possible dehiscence of the zonules inferiorly. The implant's decentration caused spherical aberrations, with prismatic effects that could be responsible for the induced myopia in this case.

A lens exchange was necessary to correct the condition and restore emmetropia to the left eye.

Induced myopia due to inadvertent use of a stronger power IOL

The determination of IOL power is done preoperatively, and the proper IOL is selected at the time of surgery. The surgeon should always double check the IOL power prior to using it, since errors can occur.

Case presentation:
This 68-year-old lady was scheduled for surgery on her left eye. The preoperative measurements were as follows:

	OD	OS
Vision	20/50	20/80
Refraction	Plano-0.50 X 180	Plano-0.50 X 160
Axial length	24.15 mm	24.40 mm
K readings	43.00/43.50 D	43.00/43.25 D

Aiming for emmetropia the surgeon planned to use an 18.5 diopter posterior chamber lens. During the surgery, a problem happened with the sterilizer requiring the presence of the head nurse outside the operating room. Ten minutes later the nurse rushed back to the operating room just in time to hand the scrub nurse the implant. She misread the operative schedule and handed the scrub nurse a 20.0 D lens, (which was the selected implant for the next patient on the schedule). The mistake was not caught until the patient was in the recovery room.

Fortunately, in this case, the inadvertent use of the 20.0 D implant resulted in an induced myopia of only -1.00 diopter without any complaints from the patient.

3. *INDUCED HYPEROPIA*

When confronted with an unexpected postoperative hyperopia, the following steps are followed in search of the cause(9-21):

- Both eyes are remeasured to rule out an erroneous preoperative longer axial length measurement.
- The corneal power is recalculated to rule out a postoperative flattening of the cornea.
- In short eyes, the IOL formula is rechecked for accuracy.

Induced hyperopia due to longer axial length measurement

A longer axial length can occasionally be measured prior to cataract surgery. The causes of such a longer measurement are detailed in Chapter VI. The use of a longer axial length measurement in IOL power calculations will call for the use of a weaker IOL than is actually required, resulting in an induced hyperopia in the final postoperative refraction. It is the most common reason for induced hyperopia in the final postoperative refraction.

Case presentation:
A 62-year-old male was seen in consultation five months after cataract surgery due to blurring of vision in his dominant operated right eye. Preoperative records from his surgeon revealed the following parameters:

	OD	OS
Vision	20/50	20/50
Refraction	Plano	+0.25 Sph
Axial length	24.0 mm	None taken
K readings	44.75/44.75 D	44.50/44.50 D

Aiming for emmetropia, the surgeon used a 16.5 D posterior chamber lens. After surgery, unaided visual acuity was blurry and could be improved to 20/20 with +1.50 Sph.

When re-examined the following parameters were noted:

	OD	OS
Vision	20/20	20/30
Refraction	+1.50 Sph	+0.25 Sph
Axial length	23.4 mm	23.4 mm
K readings	44.75/44.75 D	44.50/44.50 D

Power for emmetropia using a plano-convex posterior chamber lens:

Shammas	17.85 D	18.15 D
SRK-T	17.98 D	18.23 D
Holladay	18.00 D	18.28 D

The 0.6 mm longer measurement in this case accounted for the induced hyperopia. The patient was referred to his surgeon for a lens exchange.

Case presentation:
This 74-year-old man was referred for consultation regarding an unexpected hyperopic refraction in his right eye after cataract surgery. Preoperative records from his surgeon revealed the following parameters:

	OD	OS
Vision	20/80	20/30
Refraction	-2.75-1.00 X 102	-1.25-0.75 X 52
Axial length	24.80 mm	25.30 mm
K readings	42.75/43.25 D	42.75/43.50 D

The patient had always been slightly myopic in both eyes. The increased myopia in the right eye was attributed to advanced nuclear sclerosis. The patient underwent an uneventful phacoemulsification of the right cataractous lens with insertion of a 16.0 diopter posterior chamber lens. Postoperatively, the patient demonstrated an unexpected hyperopia in the operated eye and repeated measurements revealed the following parameters:

	OD	OS
Vision	20/50	20/40
Refraction	+2.50-0.75 X 85	-1.25 -1.00 X 45
Axial length	23.52 mm	25.06 mm
K readings	43.00/43.25 D	43.25/43.75 D

Power for emmetropia using a plano-convex IOL:

Shammas	19.55 D	14.57 D
SRK-T	19.27 D	14.58 D
Holladay	19.49 D	14.74 D

Due to the discrepancy between the axial length measurement in both eyes a B-scan ultrasonography was performed. It revealed the presence of myopic degenerative changes in both eyes; a small staphyloma involved the macular area of the left eye, but was noted nasal to the macula in the operated right eye. The technician who measured the right eye preoperatively used a contact technique and searched for the longest measurement which was through the staphylomatous area, nasal to the macula. The immersion technique in this difficult and unusual case allowed a better alignment with the macular area and a more correct axial length reading.

The patient was referred back to his surgeon for a lens exchange. The surgeon replaced it with a 19.0 diopter implant, and vacuumed the cloudy posterior capsule. Vision recovered to 20/40 with a refraction of +0.25 -0.75 X 85.

Hypermetropization of a previously myopic eye is one of the worst errors in IOL power selection. Patients do not seem to tolerate that change, and they can become irate very quickly. If such an error occurs, it behooves the surgeon to redo all calculations and exchange the implant as soon as the error is detected.

Induced hyperopia due to postoperative corneal flattening

Postoperative corneal flattening is a rare occurrence after a routine extracapsular cataract extraction. Occasionally the corneal flattening occurs in the vertical axis, secondary to a wound gape.

Case presentation:

This 67-year-old man underwent an uneventful, planned extracapsular cataract extraction to the right eye. Preoperative examination revealed the following:

	OD	OS
Vision	20/100	20/50
Refraction	Plano	Plano
Axial length	23.41 mm	23.43 mm
K readings	43.00/43.00 D	42.75/42.75 D

Power for emmetropia with a biconvex posterior chamber lens:

Shammas	20.96 D	21.22 D
SRK-T	20.65 D	20.84 D
Holladay	20.83 D	21.06 D

Planning for emmetropia, a 21 D. was used. After surgery, vision improved to 20/30 with a refraction of +3.00-3.50 X87 with a spherical equivalence of +1.25 Sph. The induced hyperopia in this case was due to a wound gape with corneal flattening in the vertical axis. The K readings were 40.50 X 90/43.50 X 180. Repair of the wound gape corrected the condition.

Corneal flattening also has been noted as a complication of cataract surgery after radial keratotomy (22). A report of four such patients revealed an initial hyperopic shift caused by an early postoperative corneal flattening that partially regressed during the follow-up period. The corneal curvature measured with the keratometer was not very accurate for IOL power calculations; a more accurate value is derived from corneal topography or by subtracting the refractive change induced by the radial keratotomy from the patient's keratometric measurements obtained before radial keratotomy. (Chapter VII)

Induced hyperopia due to using the SRK-I formula in the short eye

The original SRK-I has been replaced by the more accurate SRK-T formula. However, a few years ago, the SRK-I formula was widely used and is still used by some surgeons. The SRK-I formula calls for the use of weaker IOL in the short eyes that results in an induced hyperopia (11).

Case presentation:
This 72-year-old man underwent an uneventful cataract surgery to the right eye. The preoperative measurements were as follows:

	OD	OS
Vision	20/100	20/100
Refraction	+5.00-0.50 X 180	+4.75-0.25 X 178
Axial length	21.67 mm	21.82 mm
K readings	42.15/43.00 D	42.25/42.75 D

Power for emmetropia using a biconvex posterior chamber lens:

Shammas	27.77 D	27.34 D
SRK-T	26.43 D	26.05 D
Holladay	27.03 D	26.63 D
SRK	25.27 D	25.00 D

Aiming for emmetropia with the SRK formula, the surgeon used a 25 diopter implant. After surgery, the visual acuity improved to 20/30 with a refraction of +2.50-0.75 X 180. In spite of the error in calculations, the patient was happy for having less hyperopia than before surgery. Six months later, the second eye underwent cataract surgery with the implantation of a 25.5 D implant that resulted in a postoperative refraction of +1.50 Sph.

Induced hyperopia due to intentional emmetropization of a myopic eye

Patients with low myopia ranging from -1.50 D to -3.00 D have a choice between becoming emmetropic or staying myopic to preserve their reading vision. Problems can arise with patients who are emmetropic in the dominant eye and myopic in the other eye. These patients use the dominant eye for distance vision and the other eye for reading. Restoring emmetropia to the myopic eye can cause patient dissatisfaction.

Case presentation:
This 59-year-old lady underwent an uneventful cataract surgery to the left eye. Preoperative examination revealed the following:

	OD	OS
Vision	20/30	20/200
Refraction	+0.25 Sph	-2.50 Sph
Axial length	23.51 mm	24.20 mm
K readings	43.00/43.00 D	43.25/43.75 D

Power for emmetropia using a biconvex posterior chamber lens:

Shammas	20.63 D	17.76 D
SRK-T	20.63 D	17.88 D
Holladay	20.52 D	17.86 D

In an effort to restore isometropia in this patient, an 18 D implant was used. Postoperatively, the visual acuity improved to 20/20 with a refraction of plano -0.50 X 07. The patient was displeased with the result. There was a lack of judgment in this case and a lack of communication with the patient. The patient should have received a stronger power IOL to keep the non-dominant left eye myopic like it had been and to preserve her unaided reading vision in this eye. Furthermore, no iseikonic lens power calculations were performed. To preserve iseikonia, the left eye should have received a 21.35 D. lens with a resultant refraction of -2.75 Sph. The patient, reluctantly, now wears glasses for reading.

Induced hyperopia due to the inadvertent use of a weaker IOL

The determination of the IOL power is done preoperatively and the proper IOL is selected at the time of surgery. Human errors, due to miscommunication, are possible which can result in the use of an erroneous IOL.

Case presentation:
A surgeon scheduled two cataract surgeries at a new facility. Both patients were ethnic and did not speak the language. Both patients arrived at the facility at the same time, and were taken to the holding room at the same time. Due to miscommunication between nurses and the patients and between the nurses and the surgeon, the second patient on the schedule was ushered to the operating room and had her surgery first. Following his own schedule, the surgeon used an 18.5 diopter lens instead of the intended 20.5 diopters. The preoperative readings were as follows:

	OD	OS
Vision	20/40	20/70
Refraction	-0.25 Sph	-0.50-0.75 X 68
Axial length	23.67 mm	23.70 mm
K readings	42.75/42.75 D	42.75/43.75 D

The error was not noted by the nurses until later in the day, when they were completing the paperwork.

The patient's eye healed uneventfully and achieved a 20/25 vision with a refraction of +1.50-0.50 X 15.

Induced hyperopia due to the backward insertion of plano-convex implant

A plano-convex implant has most of its power at its convex side which conventionally is aimed anteriorly towards the cornea. The backward insertion of a plano-convex implant will move the optical center more posteriorly, resulting in an induced hyperopia.

Case presentation:

A 58-year-old myopic man underwent an uneventful phacoemulsification cataract surgery to his dominant left eye with a primary insertion of a 17.5 diopter implant resulting in an excellent uncorrected distance vision in that eye. Three months later, he was scheduled for surgery to his right eye. Preoperative examination revealedthe following:

	OD	OS
Vision	20/200	20/20
Refraction	-5.00-1.50 X 85	+0.25-0.75 X 82
Axial length	24.30 mm	24.12 mm
K readings	43.50/45.00 D	44.00/44.50 D

Power for emmetropia using a plano-convex posterior chamber lens was 15.74 D. In an effort to leave him slightly more myopic in this non-dominant eye, it was decided to use a 17.50 D. implant. However, during surgery, it was noted that the posterior capsule was ruptured with no vitreous loss and that the pupil had constricted to approximately 5 mm. The surgeon decided to insert an anterior chamber lens instead. A review of the calculations showed that a 14.58 D. implant was needed for emmetropia and a 16.0 D. implant was used to keep the eye myopic by approximately -1.00 D. Inadvertently, the implant was inserted backwards with the convex surface aiming posteriorly. Postoperatively, the eye healed uneventfully with a refraction of +1.25 -1.50 X 122 (spherical equivalence of +0.50) and a distant corrected vision of 20/20.

In this case, the anterior chamber depth constant, which is closely associated with the distance between the corneal apex and the implant's optical center, had been changed from the usual 2.9 mm for an anterior chamber implant (23) to approximately 3.9 mm. This resulted in an induced hyperopia of +1.50 D. Luckily, the patient was happy with the postoperative result.

4. INDUCED ANISEIKONIA

Aniseikonia, or the difference in the image size between the two eyes, can be induced after cataract surgery and cause patient discomfort. Unfortunately, many surgeons only calculate the IOL power for emmetropia without too much consideration for the IOL power for iseikonia. This becomes critical in cases where the cataract is unilateral and where the normal fellow eye is highly myopic, hyperopic and/or in the presence of an axial length discrepancy of over 0.3 mm between the two eyes (23,24).

Induced aniseikonia when both eyes are highly myopic

Patients with high myopia requiring cataract surgery often want to be less myopic after the surgery. However, in the presence of a unilateral cataract, the task becomes more difficult with high risks of inducing aniseikonia.

Case presentation:
A 71-year-old patient had a cataract in the left eye with decreased vision. The right eye only showed incipient cataract changes. The preoperative measurements were as follows:

	OD	OS
Vision	20/40	20/200
Refraction	-6.50 Sph	-6.50 -0.75 X 84
Axial length	26.85 mm	26.90 mm
K readings	45.25/45.25D	45.50/45.50 D
IOL for emmetropia	8.04 D	7.58 D

After detailed discussion with the patient, it was decided to aim for a postoperative myopia of around -2.50 Sph. to avoid a large anisometropia and to allow the patient to read comfortably without glasses. Postoperatively, the patient achieved 20/20 corrected vision, but kept complaining of severe discomfort. Calculations were repeated with special consideration to the iseikonic IOL power in the left eye (see Chapter III):

IOL power for emmetropia:	7.58 D
IOL power for iseikonia:	17.00 D
Iseikonic refraction:	-7.25 Sph.
5% iseikonic tolerance range:	-4.50 Sph. to -10.00 Sph.
Final refraction:	-2.25 -0.50 X 90
Induced aniseikonia	8.5%

A few months later, the cataract in the right eye progressed further necessitating surgery. Iseikonia and isometropia were reached by making the right eye myopic by around -1.75 Sph.

Induced aniseikonia when one eye only is myopic

Case Presentation:
A 62-year-old patient was examined after surgery to her left eye due to continued postoperative discomfort and asthenopia. Preoperative measurements, obtained from the operating surgeon, revealed the following:

	OD	OS
Vision	20/30	20/100
Refraction	P1-0.50 X 90	-2.75 Sph.
K readings	44.00/44.00 D	43.50/44.00 D
Axial length	23.43 mm	24.60 mm

Aiming for emmetropia in this left eye, the surgeon implanted a 16.5 D posterior chamber lens. Postoperatively the visual acuity improved to 20/25 with a refraction of +0.25 -0.50 X 85.

Calculations were repeated with special consideration to the iseikonic IOL power in the left eye:

IOL power for emmetropia:	16.22 D
IOL power for iseikonia:	22.44 D
Iseikonic refraction:	-4.75 Sph.
5% iseikonic tolerance range:	-2.00 to -7.50 Sph.
Final refraction:	+0.25 -0.50 X 85
Induced aniseikonia:	8%

The left eye should have been kept nearsighted by around -2.50 diopters to avoid aniseikonia and to preserve the unaided reading vision in the left, non-dominant eye.

Induced aniseikonia in hyperopic eyes

Hyperopes wear glasses for distance and reading and it is not unusual to aim for postoperative emmetropia in these cases. Also, these patients rarely suffer from postoperative aniseikonia due to the induced myopia required to obtain equality in the image size between the two eyes. The IOL power for iseikonia has to be considered in cases of unilateral cataract associated with a discrepancy in axial length between the two eyes.

Case presentation:
A 68-year-old patient had a cataract in the left eye with decreased vision. The preoperative measurement revealed the following parameters:

	OD	OS
Vision	20/40	20/80
Refraction	+3.75-0.50 X 97	+4.25-0.50 X 84
Axial length	23.52 mm	23.04 mm
K readings	42.00/42.00 D	42.50/43.00 D

Aiming for near emmetropia, a 22 D. posterior chamber lens was used resulting in an uncorrected vision of 20/30 and a corrected vision of 20/25 with a refraction of +0.75 Sph. The patient kept complaining of severe postoperative asthenopia, blaming it on possible undisclosed operative complications. Calculations were repeated with special consideration to the iseikonic IOL power in the left eye:

IOL power for emmetropia:	22.55 D.
IOL power for iseikonia:	17.47 D.
Iseikonic refraction:	+3.75 Sph.

5% iseikonic tolerance range:	+1.00 to +6.50 Sph.
Final refraction:	+0.75 Sph.
Induced aniseikonia:	5.5%

It took six months and multiple sessions of endless discussions to explain to her about aniseikonia and the chances for emmetropia. She finally consented to have the right eye operated upon, reached emmetropia and was cured of her asthenopia.

REFERENCES - CHAPTER X

1. Binkhorst, RD: Pitfalls in the determination of intraocular lens power without ultrasound, Ophthalmic Surg. 76:69-82, 1976.

2. Clevenger, CE: Clinical prediction versus ultrasound measurement of IOL power, Am. Intraocular Implant Soc. J. 4:222-224, 1978.

3. Kraff, MD, Sanders, DR, and Lieberman, HL: Determination of intraocular lens power: a comparison with and without ultrasound, Ophthalmic Surg. 9:81-84, 1978.

4. Olson, RJ: Intraocular lens power calculations: an extra edge or expensive waste? Arch. Ophthal. 105:1035-1036, 1987.

5. Salz, JJ, Reader, AL,III: Lens implant exchanges for incorrect power: results of an informal survey. J. Cataract Refractive Surgery. 14:221-224, 1988.

6. Shammas, HJ: The 9 diopter surprise revisited. Journal of Cataract and Refractive Surgery. 14:580, 1988.

7. Shammas, HJ: Axial length measurements and IOL power calculations in microphthalmic eyes. In "Ophthalmic Ultrasonography". Proceedings of the 12th SIDUO Congress. R Sampaolesi, Editor. Kluwer Academic Publishers. Documenta Ophthalmologica. 53:145-148, 1990.

8. Shammas, HJ and Milkie, CF: Mature cataracts in eyes with unilateral axial myopia. Journal Cataract Refractive Surgery. 15:308-311, 1989.

9. Cravy, T: Using the intraocular lens refraction factor to improve refractive prediction accuracy. J. Cataract Refractive Surg. 15:519-525, 1989.

10. Hillman, JS: Intraocular lens power calculation for emmetropia: a clinical study. Br. J. Ophthal. 66:53-56, 1982.

11. Hoffer, KJ: Intraocular lens calculations: the problem of the short eye. Ophthal. Surg. 12:269-272, 1981.

12. Holladay, JT, Praeger, TC, Long, SA, Koester, CJ: Determining intraocular lens power within the eye. AIOIS J. 11:353-363, 1985.

13. Holladay, JT, Rubin, ML: Avoiding refractive problems in cataract surgery. Surv. Ophthalmology. 32:357-360, 1988.

14. Hoffer, KJ: Accuracy of ultrasound intraocular lens calculation. Arch. Ophthal. 99:1819-1823, 1981.

15. Huber, C: Effectiveness of intraocular lens calculation in high ametropia. J. Cataract Refractive Surg. 15:667-672, 1989.

16. Olsen, T: Sources of error in intraocular lens power calculations, J. Cat. Refr. Surg. 18:125-129, 1992.

17. Olson, RJ, Kolodner, H, Kaufman, HE: The optical quality of currently manufactured intraocular lenses. Am. J. Ophthalmol. 88:548-555, 1979.

18. Phillips, P, Perez-Emmanuelli, J, Rosskothen, HD, Koester, CJ: Measurement of intraocular lens decentration and tilt in vivo. J. Cataract and Refractive Surgery. 14:129-135, 1988.

19. Richards, SC, Olson, RJ, Richards, WL: Factors associated with poor predictability by intraocular lens calculation formulas. Arch. Ophthal. 103:515-518, 1985.

20. Shammas, HJ: Accuracy of lens power calculations with biconvex and meniscus intraocular lenses. American Journal of Ophthalmology. 106:613-615, 1988.

21. Shammas, HJ: Atlas of Ophthalmic Ultrasonography and Biometry. pp. 273-308. The C.V. Mosby Co., St. Louis, 1984.

22. Koch, DD, Liu, JF, Hyde, LL, Rock, RL, Emery, J: Refractive complications of cataract surgery after radial keratotomy. American J. Ophthalmology 108:676-682, 1989.

23. Shammas, HJF: Postoperative anterior chamber depth for anterior chamber lenses, Am. Intraocular Implant Soc.J. 6:153-155, 1980.

24. Shammas, HJ: Integrated iseikonic equation for intraocular lens power calculations. J. Cat. Refract. Surg. 20:74-77, 1994.

Index

A

B

C

D

E

F

G

H